ROTHO
AUTHOR

J.M.Day
1975

Soil Microbiology

Soil Microbiology

Edited by N. Walker, Ph.D., F.R.I.C.
Rothamsted Experimental Station, Harpenden, Herts.

BUTTERWORTHS
London and Boston

THE BUTTERWORTH GROUP

ENGLAND
Butterworth & Co (Publishers) Ltd
London: 88 Kingsway, WC2B 6AB

AUSTRALIA
Butterworths Pty Ltd
Sydney: 586 Pacific Highway, NSW 2067
Melbourne: 343 Little Collins Street, 3000
Brisbane: 240 Queen Street, 4000

CANADA
Butterworth & Co (Canada) Ltd
Toronto: 2265 Midland Avenue,
Scarborough, Ontario, M1P 4S1

NEW ZEALAND
Butterworths of New Zealand Ltd
Wellington: 26–28 Waring Taylor Street, 1

SOUTH AFRICA
Butterworth & Co (South Africa) (Pty) Ltd
Durban: 152–154 Gale Street

USA
Butterworths
161 Ash Street,
Reading, Boston, Mass. 01867

First published 1975

© Butterworth & Co (Publishers) Ltd, 1975
ISBN 0 408 70670 8

Printed in England by
The Whitefriars Press Ltd, London and Tonbridge

Foreword

SIR GERARD THORNTON, F.R.S.

As the oldest member of the Soil Microbiology Department at Rothamsted Experimental Station, I wish to commend this collection of essays written by present and former members to all those interested in this field. I hope it will be found stimulating and useful to students and others concerned with soil micro-organisms.

Preface

This book is a collection of chapters contributed, with one exception, by present and former members of the Soil Microbiology Department of Rothamsted Experimental Station. Systematic studies on soil micro-organisms have been in progress at Rothamsted for well over 50 years, although the earliest work was that done by R. Warington about 90 years ago. The most quoted early investigations must be those by E. J. Russell and H. B. Hutchinson on the partial sterilisation of soil by antiseptics just before the 1914–1918 war. Since then microbiological work has been carried on un-interruptedly, first in the laboratories for protozoology and soil bacteriology and then from 1940 onwards in the combined Soil Microbiology Department.

At the suggestion of the publishers, the present volume has been prepared as a collaborative effort by present and past members of the Department together with a contribution by Dr. Powlson of the Pedology Department on the important topic of the partial sterilisation of soil. The chapters are presented mainly in the form of essays surveying varied research topics to which the individual authors themselves have contributed.

Although not intended as a textbook, it is hoped that students and others may find the material useful reading and that it may serve as an introduction to different facets of soil microbiological research.

N. Walker

Contributors

The following members of the Soil Microbiology Department of Rothamsted Experimental Station, Harpenden:

Margaret E. Brown, PhD
P. J. Dart, PhD
J. M. Day, PhD
D. Hayman, PhD
Christine M. Hepper, MSc
Barbara Mosse, PhD
P. S. Nutman, PhD, FRS
F. A. Skinner, PhD
N. Walker, PhD

together with:

J. F. Darbyshire, PhD
Department of Microbiology,
Macaulay Institute for Soil Research,
Aberdeen

R. M. Jackson, PhD
Department of Microbiology,
University of Surrey

D. S. Powlson, PhD
Department of Pedology,
Rothamsted Experimental Station,
Harpenden

Contents

1 Anaerobic Bacteria and their Activities in Soil

F. A. SKINNER

1.1 Introduction

A variable proportion of the microbial population in soil consists of anaerobic micro-organisms, some of which grow only in the absence of oxygen (obligate anaerobes); others can grow with oxygen, or anaerobically when supplied with suitable nutrients (facultative anaerobes). Although the existence of such organisms has been known since the beginning of soil microbiology, the extreme diversity of this population has only gradually become revealed by the continued development of cultural methods.

Some groups of soil anaerobes, such as those that fix nitrogen, decompose cellulose, form methane, or reduce sulphate to sulphide or nitrate to gaseous nitrogen, are of especial interest to the agricultural microbiologist. There are many other anaerobes which may, nevertheless, effect important transformations of nutrients in soils. Many of these organisms have been studied in laboratory culture, but little is known about their behaviour in constantly changing soil conditions or about their possible importance in influencing soil fertility and structure.

One main cause of anaerobic conditions is physical displacement of the soil air by water through which oxygen diffuses slowly: such waterlogging, if prolonged, favours proliferation of anaerobes and leads to strongly reducing conditions. A second cause is depletion of oxygen in the soil air by the respiration of soil micro-organisms at a rate faster than that at which atmospheric oxygen can diffuse to the sites of metabolic activity. Recent changes in farming practice—such as the use of heavier machinery, which compacts the soil, and the application of large amounts of manure slurry which prevents access of air to the soil surface—are likely to increase the chance of some soils staying anaerobic for long periods.

1

Interest in the phenomenon of anaerobiosis in relation to soil structure and fertility was indicated in a recent report.[1] In this the importance of a good soil structure permitting of access of adequate air to plant roots was, rightly, stressed. The onset of anaerobic conditions was generally held to be undesirable because they would either prevent improvement in soil structure or cause structure already developed under aerobic conditions to deteriorate. The report also mentions that microbial activity in the absence of oxygen could lead quickly to the formation of substances such as ammonia, sulphide, ferrous iron, phosphine, methane and ethylene which may damage higher plants. Although this may be broadly true, the report does not indicate an appreciation of the levels of reducing intensity at which these compounds are formed. Thus, ammonia is formed by the mineralisation of nitrogen compounds under aerobic as well as anaerobic conditions, though more rapidly when anaerobic.[43] Sulphides, ferrous iron and methane are certainly produced only under strongly reducing conditions, but ethylene can be formed before the redox potential falls too low—for example, when nitrate is present. The conditions under which phosphine is formed, presumably by microbial action, are not yet understood.[40] The absence of oxygen is not the sole factor that determines which anaerobes will flourish in the soil; growth will depend greatly on the reducing intensity of the environment, on the availability of nutrients and on physical factors such as temperature and water content.

An attempt is made here to summarise much of what is known about the anaerobic soil bacteria and to discuss their importance to soil fertility and structure. Special attention is given to those anaerobes concerned with transformations of nitrogen, and to investigations that have been made at Rothamsted.

1.2 Facultative anaerobes

The facultative anaerobes, which can grow in the absence of oxygen provided suitable nutrients are present, do not constitute a homogeneous group. Two basic types of facultative anaerobe can be recognised: those able to use nitrate or some other nitrogen oxide in place of elemental oxygen, as the terminal hydrogen acceptor in the respiratory pathway, and those which derive energy either by oxidative respiration in the presence of oxygen or by fermentation of organic substrates in its absence, rather in the manner of yeasts. Some bacteria, e.g. *Klebsiella (Aerobacter) aerogenes,* possess both types of anaerobic respiratory mechanism.

The reduction of nitrate

The ability to reduce nitrate to nitrite is possessed by many bacteria. Of soil species in the Pseudomonadales, Eubacteriales and Actinomycetales, as listed in *Bergey's Manual*,[8] about 40, 50 and 60%, respectively, have this ability; these species occur in at least 23 genera. This energy-yielding mechanism, a dissimilatory process, is known as 'nitrate respiration' to distinguish it from assimilatory processes in which the nitrate is reduced before being incorporated into cell substance.

Nitrate reduction is an efficient process with respect to the energy yielded to the cell. According to McCarty,[27] the mean yield of energy for transferring an electron from an organic electron donor to nitrate is *c*. 18 kcal per gram equivalent, 67.9% of the energy yielded by a similar transfer to oxygen.

Nitrate can be reduced further to gaseous nitrogen by some bacteria. This process, called 'denitrification' by Gayon and Dupetit,[20] who first studied it, is effected by only a few species but these are widely distributed in soils and are often numerous. About eight genera of soil bacteria contain species able to denitrify, those in *Pseudomonas, Micrococcus* and *Bacillus* being especially well known. In a few species nitrous oxide, rather than nitrogen, is the final product of denitrification. The ability to reduce nitrite, but not nitrate, to gaseous nitrogen is rare.

Nitrate respiration depends on the presence of reductases for the different nitrogenous oxides, and associated electron transport compounds: such systems have been greatly studied in recent years and the literature has been reviewed extensively by Payne.[31] These respiratory enzyme systems tend to be absent from cells growing in strongly aerobic conditions but develop quickly when the oxygen tension falls. Lack of oxygen is the most important factor influencing the synthesis of nitrogenous oxide reductases in all species with nitrate or complete denitrifying respiration. For example, when *Pseudomonas denitrificans* was made anaerobic in the absence of nitrogenous oxides, synthesis of nitrous oxide reductase began and reached a maximum level in 3 h; nitrite and nitric oxide reductase were also formed, but to a lesser extent, in this time. Thus, there was no advantage in having nitrate present in order to ensure synthesis of the enzymes.

The importance of maintaining anaerobic conditions for the formation of denitrification enzymes is now generally recognised, but workers have not always agreed about the amount of aeration

3

that can be tolerated by denitrifiers. Collins[13] in studies on *Ps. aeruginosa* growing in 0.2% nitrate broth with continuous shaking observed that denitrification, presumably aerobic, occurred in some cultures. It was found that the ability of the organism to develop the denitrifying enzymes was related to the shape of the culture flasks; those flasks which restricted aeration allowed the denitrifying systems to develop. The use by other workers of shallow layers of undisturbed culture medium or deep medium aerated by bubbling air through it probably failed to provide enough oxygen for the rapidly growing cells and so promoted the synthesis of nitrate respiration systems. It seems clear that the rate of synthesis of the denitrifying enzymes in all bacteria capable of denitrifying respiration[31] is inversely related to the availability of oxygen.

A supply of organic compounds to act as electron donors is essential for nitrogen oxide respiration. Usually a wide range of compounds will serve for this purpose but it must not be assumed that the nature of these substances is without importance. Thus, the addition of short-chain fatty and related acids to media containing nitrate increased the nitrate reductase activity of *Mycobacterium tuberculosis*. Some species that were thought to lack the ability to reduce nitrate could do so when given the right electron donor; e.g. *Myco. scrofulaceum* reduced nitrate in cultures containing lactic acid, but not in media without it.[6]

Fermentative facultative anaerobes

The ability to derive energy by anaerobic fermentation is possessed by a wide range of micro-organisms, many of which occur in soil. These can be detected by cultivation in media that contain fermentable carbohydrate but no nitrate, anaerobic incubation or deep culture media being used. In Rothamsted soils facultative organisms of this class, predominantly bacteria and yeasts, often form about 10% of the total aerobic count on the same medium. The significance of this population to soil fertility is largely unknown.

1.3 Obligate anaerobes

Spore-formers

Most obligate anaerobes of the soil are spore-forming bacteria of the genus *Clostridium:* these occur in numbers generally within the range

10^4–10^5 per gram of soil in temperate regions. The population of soil clostridia comprises many species several of which are pathogenic to man or animals. *Clostridium welchii*, the most important causative organism of gas gangrene in man, is widely distributed and usually abundant in soils; up to 10^5 spores per gram of soil have been found. Other pathogenic clostridia, such as *Cl. oedematiens* Type A and *Cl. septicum*, are also fairly common. Non-pathogenic proteolytic clostridia also abound in soils; *Cl. sporogenes* is probably the commonest and is often associated with pathogenic types in gas-gangrenous wounds. Many other less common species are found in such lesions and in soils.[44]

Soil clostridia vary greatly in their ability to grow in the presence of oxygen. Some species, such as *Cl. sporogenes*, *Cl. oedematiens* and *Cl. septicum*, are strict anaerobes; others, such as *Cl. bifermentans*, *Cl. histolyticum* and *Cl. welchii*, are less exacting and can grow with a trace of oxygen present. *Clostridium tetani*, the causative organism of tetanus, is widely distributed in soils though not abundant: it is a strict anaerobe but not fastidious in its nutrient requirements.

Soil clostridia may be counted with a selective medium and procedure such as that of Gibbs and Freame[21]; pasteurised and un-pasteurised inocula are used to give estimates of spores and total cells (spores + vegetative cells), respectively. It is a common experience to find that these two estimates agree closely, especially when the soil sampled is dry, which indicates that the clostridia are present as spores. In a survey of 25 soils from Rothamsted, Suffolk and Oxfordshire the numbers of clostridial spores ranged from 4.3×10^3 per gram of dry soil for Rothamsted Broadbalk (plot 19) to 6.3×10^5 per gram of dry soil for Broadbalk (plot 22). Only six of the soils examined contained more than 10^5 spores per gram. Numbers were generally low, probably because the soils were sampled during a prolonged spell of dry weather.[35] *Clostridium botulinum*, a toxin-producing organism of great importance in the food industry, was not detected in any of these 25 soils but it would grow to some extent when inoculated into a moist rich soil (Roberts, personal communication).

Garcia and McKay[19] studied the growth and survival of *Cl. septicum* in soil and found that a limited amount of growth occurred in experiments made with soils at 60% of water-holding capacity incubated in nitrogen at 37°C. In sterile chernozem soil with and without added glucose or manure (both used at 1%) numbers of *Cl. septicum* were highest after 7 days; thereafter smaller numbers were found and, after 4 weeks, the numbers in all treatments approximated to those at the start of the experiment (5×10^5 per gram). This increase at 7 days was not great, a twofold increase being given by glucose or manure used separately; the increase was accompanied by a fall in the proportion of viable spores to *c.* 30% of the count in the amended soils. Sporulation increased again after 2 weeks; at 30 days the proportion of spores approached that at the start of the experiment—*c.* 70% (in sterile soil). A similar result was noted in fresh unsterile soil.

The occurrence of so many spores encourages speculation about the activity of these organisms and their ability to proliferate when conditions are anaerobic. Many species of clostridia occur in the intestinal tract and faeces of man and animals; *Cl. welchii* and *Cl. sporogenes* are abundant there. Moreover, most clostridia grow well or optimally at 37°C, a temperature rarely attained in soils except on the surface, where aeration and dryness would preclude their growth. Possibly they are adapted primarily to the intestinal environment but can survive well in soil. On the other hand, clostridia are found in soils far from human habitation or regions likely to be contaminated by animal droppings. Thus, it is arguable that clostridia really are soil organisms and that a growth temperature optimum of 37°C is fortuitous. Also, our awareness of so many clostridia able to grow optimally at the temperature of the human body may reflect the habits of microbiologists, who frequently cultivate organisms at temperatures well above those likely in nature and so select organisms that grow at high temperatures. This view receives support from a consideration of the genus *Bacillus,* the aerobic spore-forming bacteria. In this genus, of 16 soil species listed[8], seven grow optimally or well at 37°C; nine grow similarly at 30°C; five species grow well at both temperatures. These aerobic forms, which are not typical intestinal organisms, nevertheless grow at temperatures higher than usually occur in soils.

The clostridia considered so far are especially important from the medical or public health standpoint but some also have useful functions in soil. For example, *Cl. sporogenes* is proteolytic and probably responsible for much decomposition of animal tissue when conditions are anaerobic. Proteolytic properties are common among the species already listed, but there are many soil clostridia with very limited proteolytic ability but which can ferment carbohydrates vigorously; these are active in decomposing plant remains. These saccharolytic forms, typified by *Cl. butyricum* and *Cl. pasteurianum,* are of particular interest to the soil microbiologist because many strains can fix atmospheric nitrogen.

Clostridia able to decompose cellulose anaerobically are common in soils but not often numerous unless anaerobiosis is strict and prolonged. Anaerobic enrichment cultures[39], using cellulose as carbon source, quickly reveal the presence of cellulolytic spore-forming anaerobes in soils, but they can be isolated only under strict anaerobiosis with a low redox potential (E_h). In pure culture these stringent conditions need to be maintained if growth is to continue. Growth does not occur at the comparatively high E_h prevailing when nitrate is present. Thus, anaerobic cellulose decomposition in soil is probable only when anaerobiosis is prolonged, as, for example, by waterlogging, in the absence of nitrate. For this reason one cannot regard these organisms as being of prime importance in cellulose decomposition in soil.

Clostridia that decompose pectin have been known for a long time in the retting of flax. According to Emtsev, Razvozhevskaya and Dzadzamiya[18], who have studied the distribution of pectinolytic forms in different soil types of the USSR, some of them fix nitrogen. Chitinolytic clostridia have been isolated from sea-shore mud,[5] and probably they have a wide distribution in soil where chitin occurs in fungus and insect residues.

Nitrogen-fixing clostridia

Many strains of saccharolytic clostridia, especially of *Cl. pasteurianum,* fix atmospheric nitrogen when provided with a fermentable carbon source and very little combined nitrogen. Although members of the aerobic genus *Azotobacter* fix nitrogen efficiently in culture, they are relatively scarce in soils[28, 29] and so cannot play a major part in the accumulation of fixed nitrogen. Nitrogen-fixing clostridia are more abundant: this fact, and the realisation that nitrogen fixation is essentially an anaerobic process, justify a detailed appraisal of the role of these clostridia in the soil economy.

When a nitrogen-free mineral salt solution (e.g. glucose, 10 g; K_2HPO_4, 1 g; $MgSO_4 \cdot 7H_2O$, 0.2 g; water, 1 litre) containing a little chalk is inoculated with fresh soil and incubated (18–20°C) with a slow current of nitrogen passing through it, fermentation begins in a few days. Nitrogen-fixing clostridia, often *Cl. pasteurianum,* can be isolated from this enrichment culture by the usual methods of anaerobic bacteriology.

It is easy to grow these clostridia in such enrichment cultures because the heavy soil inoculum provides essential growth factors and minerals, but in making pure cultures these essential nutrients must be supplied separately. Iron and molybdenum are needed in traces because they are components of the nitrogenase (nitrogen-fixing enzyme) system. All clostridia need organic growth factors, those for nitrogen fixers being conveniently supplied by yeast, soil or potato extracts.

In his original description of *Cl. pasteurianum* Winogradsky[45] drew attention to the fact that cells in which spores were developing stained dark brown or violet–brown with iodine. This property has been used to identify this organism microscopically and forms the basis of a counting method devised by Jensen.[25] Test tubes containing pieces of potato, chalk and water are sterilised, inoculated with appropriate soil dilutions, incubated and examined after a few days for cells staining with iodine; such tubes, regarded as containing nitrogen-fixing clostridia, usually show production of gas and have a butanol or butyric acid odour. Methods for counting and isolating these organisms have been reviewed in detail.[41]

With the potato tube method Meiklejohn[29] counted nitrogen-fixing clostridia in soils from plots of Broadbalk field, Rothamsted. Numbers ranged from 10^3 to 10^5 per gram; azotobacter cells were fewer than 10^3 per gram of soil.

The method probably overestimates the population of nitrogen-fixing clostridia. Experiments on the suitability of different nitrogen-deficient liquid media for counting these anaerobes showed that staining with iodine was an unreliable method.[34] Moreover, with Rothamsted soils the potato tube method usually gives higher, sometimes about 100-fold, estimates of presumptive nitrogen-fixing anaerobes than parallel estimates of the total clostridial population, including all proteolytic types, given by the method of Gibbs and Freame[21], a method of known efficacy for estimating clostridia in foods. Experiments with pure cultures have shown that strains of *Cl. butyricum,* some of which fix nitrogen, usually blacken the medium of Gibbs and Freame[21] but that strains of *Cl. pasteurianum* do so less reliably.[35]

Despite these discrepancies in counting methods, the opinion that anaerobes are active nitrogen-fixing agents in Broadbalk soil, and probably in other soils as well, is supported by laboratory observations of nitrogen fixation by soil from Broadbalk incubated anaerobically with straw.[3] Further support for this view comes from Day *et al.,*[17] who showed that nitrogenase activity, as measured by the acetylene-reduction technique, increased with increasing soil water content. This suggests that nitrogenase activity and, presumably,

8

growth of the organisms concerned are favoured by restricted aeration. So far, this activity has not been related to the numbers and types of nitrogen-fixing micro-organisms; this must depend on further improvements being made in the methods for their enumeration and identification.

Other obligate anaerobes

Two other groups of obligate soil anaerobes—the sulphate-reducing and the methane-forming bacteria—merit special mention because of their unusual physiology and their potential importance when anaerobiosis is prolonged.

Unlike facultative anaerobes with nitrate respiration, the dissimilatory sulphate-reducing bacteria are obligate anaerobes and have no other energy-yielding mechanism. These organisms occur widely in nature and flourish in wet or waterlogged soils and permanent muds provided that oxygen is absent and sulphate (the electron acceptor) and organic matter (to supply electron donors) are available. Although sulphate reducers abound when conditions are favourable, few species have been recognised. There are two genera: non-sporulating bacteria are placed in *Desulfovibrio*[33] and spore-formers in *Desulfotomaculum*.[12] Despite an over-all similarity in possessing sulphate respiration, there are large differences between members of these genera; thus, the desulfovibrios possess cytochrome c_3 and desulfoviridin as components of the respiratory system, whereas the spore formers do not, and there is no close antigenic relationship between the two groups.

The importance of sulphate-reducing bacteria lies in their ability to produce sulphide which corrodes iron pipes and other structures that are buried in anaerobic soil. The corrosion process is partly microbial and partly electrolytic. Areas of iron in which the bacteria produce sulphide becomes anodic to those parts without sulphide; hydrogen then accumulates at the cathodic areas, and if this is removed continually by micro-organisms able to oxidise hydrogen, corrosion of the iron will progress.[7]

The microbial production of sulphide may also have some agricultural significance because it can be toxic to plant roots.[14]

The methane-forming bacteria require anaerobiosis at least as strict as that needed by the cellulose decomposers. Only small amounts of methane are likely to be formed from natural or cultivated soils. The evolution of appreciable amounts of methane

9

occurs only in exceptional circumstances where accident or mis-management has allowed too much organic matter to accumulate in soil that is liable to be too wet because of impeded drainage.[40] Methane can also be formed in soils covered with heavy layers of manure slurry (Burford, personal communication).

1.4 Anaerobes in relation to soil structure and function

Early work on anaerobic soil conditions at Rothamsted

Until recently, comparatively little interest was taken in the nature and activities of the soil anaerobes at Rothamsted, though some potentially important work was undertaken just before World War I. During an investigation on the fate of nitrogen supplied in fertilisers, it was noticed that on cropped land which had been supplied generously with farmyard manure there was a loss of nitrogen that could not be accounted for in the crop or as nitrate leached out in the drainage water. A loss of gaseous nitrogen was suspected, but analysis of soil gases over a period of at least 2 years failed to demonstrate such a loss. It was thought that denitrification, which was known but not well understood at that time, might release gaseous nitrogen from the soil at too slow a rate to be detected by the methods then available.

Studies of manure heaps by Russell and Richards[38] showed that nitrate was formed on the outside of the heap, which was aerobic, but, if the heap received enough water to wash the nitrate into it but not enough to leach out the nitrate, there was a considerable loss of gaseous nitrogen. Thus, a correlation between decomposition of nitrate and restricted aeration was established. This is readily intelligible in the light of modern knowledge: we would explain it by saying that nitrate was formed by nitrification at the surface of the heap and that any nitrate passing inside would be used in nitrate respiration by facultative anaerobes living there.

Russell regarded this work on soil gases and that on denitrification in manure heaps as providing an explanation for the loss of gaseous nitrogen from soil. Investigations on the changes in soil atmosphere at 6 in depth[37] had revealed that air in the soil spaces differed only slightly from atmospheric air in oxygen content, though there was rather more carbon dioxide. In addition to this air in the pore spaces, they found a second atmosphere, dissolved in soil water and absorbed by colloids, which contained typically little or no oxygen and much carbon dioxide. The existence of such an atmosphere in close association with soil solids indicated that soil micro-organisms were using up oxygen faster than it could diffuse to the regions of microbial activity. Russell likened the soil crumb or aggregate containing organic matter to a miniature manure heap in that it was surrounded by gas similar in composition to air (*c.* 20% O_2), and so nitrate production could go on at the surface. The inside of the aggregate would be anaerobic and would favour denitrification whenever soil moisture allowed of diffusion of nitrate to the interior.

Denitrification is undesirable from an agricultural viewpoint because it causes the loss of valuable nitrogen from the soil. The process is favoured by (*i*) a slightly alkaline reaction; (*ii*) warmth—although denitrification can take place in tundra or in other cold soils; (*iii*) lack of oxygen; and (*iv*) adequate organic matter to supply electron donors. Soil need not be very wet for some denitrification to take place; this is explainable in terms of the concept of the anaerobic aggregate surrounded by soil air with a near-atmospheric oxygen content.

Denitrification in waterlogged soils

According to Bell[4], a waterlogged soil usually develops a redox potential (E_h) of *c.* +200 mV and maintains this for a period during which fermentation of carbohydrate occurs and the gas evolved contains carbon dioxide and hydrogen. After a lag period the E_h falls sharply to *c.* −200 mV and the fermentation gas begins to contain methane. When nitrate is present, denitrification proceeds during the first phase and the gas contains some nitrogen: the E_h does not fall below +200 mV. When all nitrate has been decomposed, the E_h remains no longer poised at +200 mV but falls to −200 mV, which is about the threshold value for methane formation. Conditions then become stabilised so long as reserves of nutrients for the methanogenic bacteria remain.

The addition of nitrate delays the onset of highly reducing conditions and methane formation. If there is enough nitrate to ensure complete utilisation of electron donors by the denitrifiers, then methane formation will be prevented completely.

Nitrate in soil is usually regarded simply as a plant nutrient which has the disadvantage of being easily leached out in drainage water and so lost to the crop. Consequently, there have been attempts to suppress the activities of the aerobic nitrifying bacteria in order to

11

preserve mineralised nitrogen as ammonia which is retained by soil colloids against leaching. Inhibitors of nitrification such as N-Serve have been used for this purpose. However, in view of what has been stated above about the behaviour of nitrate in waterlogged soil, it is arguable that nitrate is not simply a plant nutrient with inconvenient properties but also a useful redox potential poising agent. Clearly, if nitrate is absent from soil, by accident or design, then waterlogging could be followed quickly by the development of strongly reducing conditions inimical to the growth of higher plants.

Growth of anaerobes in non-waterlogged soil

Winogradsky[46] noted that soil was usually considered to be aerobic and he questioned whether the anaerobic microflora became active when the soil was compacted, very wet or generally badly aerated. In an experiment designed to shed light on this matter he relied on microscopic examination of the soil to judge whether it was anaerobic. The presence of *Azotobacter,* which is easily recognised by its large cell size, was the criterion of aerobic conditions and the presence of spore-forming bacteria ('amylobacters'), including *Clostridium pasteurianum,* was taken as an index of anaerobiosis. Both these organisms flourish in soil poor in nitrogen, provided that some carbohydrate is present. Winogradsky took samples of one soil type with different natural water contents, added 1% of mannitol or glucose as dry powders and packed the treated soils into tubes of 5 cm diameter to give vertical soil columns 4–23 cm long. Soil with 15.5% of water (31% of the water-holding capacity) contained abundant azotobacter throughout and was therefore judged to be aerobic. Only when the water content reached *c.* 20% did the situation change markedly; spore-bearing anaerobes staining darkly with iodine were numerous to within 4–5 cm below the soil surface and an odour of butyric acid was detectable near the bottom of the column. At 23% moisture (48% of water-holding capacity) the anaerobes were growing almost at the surface, a butyric odour was noticeable at the top and the whole column was broken by gas bubbles resulting from fermentation of the carbon source added. As a result, it was concluded that the anaerobes, inactive in moderately dry soils, can quickly become active at a water content far below that required to waterlog the soil. When a critical point is reached (*c.* 40% of water-holding capacity for the soil tested) anaerobes can flourish even almost up to the surface, provided that suitable carbon sources are present. Winogradsky suggested that anaerobiosis plays a more important role in arable soils than is generally supposed, but he did not indicate what that role might be.

There is little doubt that the proliferation of anaerobes of the 'amylobacter' type was preceded by sufficient activity of aerobic organisms to cause drastic lowering of the oxygen tension in the soil pores, this activity resulting from the large amount of readily available carbon source added to the soil. Such materials as sugars are not usually added to soils and much of the organic matter normally supplied is of a more resistant type such as cellulose, which decomposes slowly and never yields a flush of easily decomposable substrate. However, there are occasions on which sugary or other readily available substrates are applied as soil dressings. Manure slurry is often applied to arable or pasture land as fertiliser and sometimes in much larger doses as a disposal procedure. Leaf protein liquor may also become available and need to be disposed of, preferably in a profitable way. It is, therefore, necessary and instructive to consider the likely results of such applications.

Application of leaf protein liquor to soil

Leaf protein[32] is produced by expressing the juice from pulped plant tissue and heating it to coagulate the dissolved protein, which is then removed by filtration. The remaining liquor or 'whey', rich in salts, sugars and nitrogen compounds, will support good growth of micro-organisms. The possibility of using mobile equipment in the field for partial extraction of protein from fodder crops destined subsequently to be dried raises the question of disposal of the whey with minimum transport costs, possibly by direct application as a fertiliser to the soil.[2] Preliminary studies in which whey made from cocksfoot grass was added to soil from Saxmundham, Suffolk—a Boulder Clay soil difficult to work—indicated an improvement in structure. When this soil was incubated at 25°C with enough liquor to give just waterlogged conditions, fermentation ensued: at 24 h the soil had expanded slightly; after 48 h the volume of soil had almost doubled as a result of copious gas formation; and it remained in this expanded condition on drying. Similar results were found when solutions of sucrose or molasses, containing comparable amounts of sugar to that in the leaf protein whey, were used instead.

Microbial changes in this system were studied by incubating 10 g portions of air-dry Saxmundham soil with 6 ml portions of water (control), 0.3% or 3% molasses solution in glass vials (25 x 50 mm). Vigorous gas production and consequent soil expansion were associated with marked increases in numbers of anaerobes when 3% molasses was used. These micro-organisms also flourished when 0.3% molasses was the substrate, although gas production and soil expansion was small. Twelve random isolates of anaerobic bacteria were taken from a 3% molasses–soil culture; all were spore-forming bacteria of the *Clostridium butyricum* or *Cl. pasteurianum* type. Ten of these isolates were inoculated separately into vials of sterile soil mixed with sterile 3% molasses solution and incubated. In every instance the same result was obtained as when unsterile soil had been used. Development of butyric odours and pH changes were also similar to those in unsterile soil. Thus, the observed changes in unsterile soil were consistent with the view that they had been brought about by saccharolytic clostridia. There was evidence from soil sections that bacterial polysaccharide lined pores in the treated soil and allowed the expanded structure to persist on drying.[36]

Application of manure slurries to soils

Traditionally, animal manure, often containing straw, was applied to the soil as a solid material by manhandling. Today, economic considerations have encouraged the use of liquid manure slurries that can be applied to the soil mechanically from tanker vehicles. This practice introduces pollution problems, including surface run-off of organic material to water courses, possible spread of pathogens and accumulation of undesirable metal ions. These problems are especially acute where the need to dispose of excess manure leads to the application of more slurry to the soil than is needed as fertiliser. The heavy deposition of slurry applies much water; this and blockage of soil spaces by particles of organic matter carried down by the water lead to restriction of aeration. The organic matter applied also favours respiration of micro-organisms and leads to the development of anaerobic conditions (Burford, personal communication). The long-term consequences of such heavy applications on soil structure and fertility have yet to be evaluated.

Other activities of anaerobes

Two other soil processes associated with the anaerobic metabolism of micro-organisms deserve mention. The first concerns the production of ethylene in concentrations likely to cause damage to plant roots. Smith and Scott Russell[42] found that when soils moistened to field capacity were incubated at room temperature in sealed containers, ethylene could be detected in the soil air. The concentration of ethylene rose for 3 days and then remained fairly steady at 1 ppm for one soil and 10 ppm for the other two. Traces of other hydrocarbons, such as ethane and propane, were also detected. Several micro-organisms able to form ethylene in culture have been reported by Lynch[26]; two yeast isolates and the fungus *Mucor hiemalis* produced abundant ethylene in pure culture.

Decomposition of a pesticide

As a general rule, organic compounds are decomposed more rapidly and completely under aerobic than under anaerobic conditions, but there is one instance of a pesticide being resistant to attack aerobically but being subject to at least partial degradation when oxygen is deficient. Guenzi and Beard[24] found that less than 1% of a dose of DDT applied to anaerobic soil was recovered after 12 weeks of incubation; in similarly treated aerobic soil *c.* 75% was recovered after the same period. Burge[11] found that the DDT was converted to some extent to DDD anaerobically. The process was inhibited by autoclaving but began again after the soil had been inoculated with a little unsterile soil. Anaerobic breakdown of DDT was accelerated by glucose or plant residues. The degradative process was highly sensitive to oxygen. The only detectable product remaining in all cases was DDD in amount equivalent to *c.* 26% of the DDT supplied initially. No other decomposition products of DDT were detected to account for the missing 74%.

1.5 Summary and conclusions

A survey of the anaerobic micro-organisms in soils reveals a wide variety of type and function. Possibly, some of the proteolytic clostridia should not be regarded as true soil forms at all. Certainly they proliferate in the intestinal tract, where their high optimum temperature requirements are satisfied. However, they can multiply to some extent in soil and they undoubtedly play an important part in the initial breakdown of proteinaceous residues.

Waterlogging, by displacing air, leads quickly to anoxia;

anaerobes become active; and, as a result of fermentation, substances such as organic acids and alcohols accumulate. Later these compounds serve as substrates for methane bacteria which become active when the redox potential becomes low enough. The presence of nitrate, by poising the redox potential at a high value, prevents methane formation. With nitrate, provided that ample organic matter is present, denitrification can proceed until the nitrate is exhausted: only then will methane evolution begin. Thus, anoxia, if prolonged, will lead first to loss of nitrogen by denitrification and ultimately to highly reducing conditions that can damage crops.

It is evident that anaerobes can flourish even when soil spaces are filled with air, provided that enough easily decomposable organic matter is present and soil temperature is high enough for micro-organisms to consume oxygen faster than it can diffuse into the soil spaces from the atmosphere. This diffusion process has been studied and found amenable to mathematical treatment.[15, 23] It is desirable that the pore spaces and channels in soil should permit of enough diffusion of air to satisfy the requirements of plant roots and aerobic micro-organisms decomposing organic matter. This condition will usually be assured, in agricultural soils, by providing adequate drainage.

The concept of soil—at least of the good arable kind—consisting of aggregates whose extremely fine pore structure impedes gaseous diffusion so effectively that their interiors are often anoxic has also been treated mathematically.[16] The presence of such locally anoxic zones probably explains why some denitrification can go on in a soil apparently well aerated.[9, 10, 22, 30] Undesirable denitrification and beneficial nitrogen fixation both require similar conditions of anoxia and supplies of organic matter. But denitrification can proceed only when nitrate or nitrite is present; nitrogen fixation will be suppressed by fixed nitrogen in many forms.

It seems evident that soil aggregates will often be anoxic: it is not evident, at our present stage of knowledge, whether the aggregates should, by appropriate soil management, be made as aerobic as possible to prevent denitrification or whether some degree of anoxia is desirable to ensure maximum aggregate stability, prevent too rapid oxidation of soil organic matter or promote nitrogen fixation. Certainly it cannot be assumed that anoxia is entirely harmful. Further work on soil anaerobes, their activities and the conditions which favour their growth is needed if these problems are to be solved.

Acknowledgements

I am indebted to Dr. T. A. Roberts, Agricultural Research Council Meat Research Institute, Langford, Bristol, for information concerning the growth of *Clostridium botulinum* in soil, and to Dr. J. R. Burford, Department of Soil Science, Reading University, for discussions on the effects of applying heavy dressings of manure slurry to soils.

REFERENCES

1. Agricultural Advisory Council. *Modern Farming and the Soil.* Report of the Agricultural Advisory Council on Soil Structure and Soil Fertility. HMSO, London (1970)
2. ARKCOLL, D. B. and DAVYS, M. N. G. The production and use of leaf protein, *The Chemical Engineer,* No. 251, July, 1971, 261–264 (1971)
3. BARROW, N. J. and JENKINSON, D. S. The effect of water-logging on fixation of nitrogen by soil incubated with straw, *Plant and Soil,* **16,** 258–262 (1962)
4. BELL, R. G. Studies on the decomposition of organic matter in flooded soil, *Soil Biology and Biochemistry,* **1,** 105–116 (1969)
5. BILLY, C. Étude d'une bactérie chitinolytique anaérobie *Clostridium chitinophilum* n.sp., *Annales de l'Institut Pasteur* (Paris), **116,** 75–82 (1969)
6. BÖNICKE, R., JUHASC, S. E. and DIEMER, U. Studies on the nitrate reductase activity of mycobacteria in the presence of fatty acids and related compounds, *American Review of Respiratory Diseases,* **102,** 507–515 (1970)
7. BOOTH, G. H. Sulphur bacteria in relation to corrosion, *Journal of Applied Bacteriology,* **27,** 174–181 (1964)
8. BREED, R. S., MURRAY, E. C. D. and SMITH, N. R. *Bergey's Manual of Determinative Bacteriology,* 7th edition. Baillière, Tindall and Cox, London (1957)
9. BREMNER, J. M. and SHAW, K. Denitrification in soil. II. Factors affecting denitrification, *Journal of Agricultural Science* (Cambridge), **51,** 40–51 (1958)
10. BROADBENT, F. E. and STOJANOVIC, B. F. The effect of partial pressure of oxygen on some soil nitrogen transformations, *Proceedings. Soil Science Society of America,* **16,** 359–363 (1952)
11. BURGE, W. D. Anerobic decomposition of DDT in soil. Acceleration by volatile components of alfalfa, *Journal of Agricultural and Food Chemistry,* **19,** 375–378 (1971)
12. CAMPBELL, L. L. and POSTGATE, J. R. Classification of the spore-forming sulfate-reducing bacteria, *Bacteriological Reviews,* **29,** 359–363 (1965)
13. COLLINS, F. M. Effect of aeration on the formation of nitrate-reducing enzymes by *Ps. aeruginosa, Nature,* **175,** 173–174 (1955)
14. CULBERT, D. L. The use of a multi-celled apparatus for anaerobic studies of flooded root systems, *Hortscience,* **7,** 29–31 (1972)
15. CURRIE, J. A. Gaseous diffusion in the aeration of aggregated soils, *Soil Science,* **92,** 40–45 (1961)
16. CURRIE, J. A. Diffusion within soil microstructure. A structural parameter for soils, *Journal of Soil Science,* **16,** 279–289 (1965)
17. DAY, J. M., HARRIS, D., DART, P. J. and VAN BERKUM, P. The Broadbalk experiment. An investigation of nitrogen gains from non-symbiotic fixation, in *IBP Synthesis Volume on Nitrogen Fixation,* Cambridge University Press (in press)

17

18. EMTSEV, V. T., RAZVOZHEVSKAYA, Z. S. and DZADZAMIYA, T. D. [Geographical distribution of soil anaerobic nitrogen-fixing *Clostridium* bacteria—in Russian, English summary], *Izvestiya Akademia nauk SSSR Ser. Biol.,* **5,** 705–712 (1969)

19. GARCIA, M. M. and McKAY, K. A. On the growth and survival of *Clostridium septicum* in soil, *Journal of Applied Bacteriology,* **32,** 362–370 (1969)

20. GAYON, V. and DUPETIT, G. Recherches sur la réduction des nitrates par les infiniment petits, *Mémoires de la Société des Sciences Physiques et Naturelles de Bordeaux,* **2,** 201–208 (1886)

21. GIBBS, B. M. and FREAME, B. Methods for the recovery of clostridia from foods, *Journal of Applied Bacteriology,* **28,** 95–111 (1965)

22. GREENWOOD, D. J. Nitrification and nitrate dissimilation in soil. II. Effect of oxygen concentration, *Plant and Soil,* **17,** 378–391 (1962)

23. GREENWOOD, D. J. Measurement of microbial metabolism in soil, in *The Ecology of Soil Bacteria:* Symposium, Liverpool, 1965 (edited by T. R. G. Gray and D. Parkinson). Liverpool University Press (1968)

24. GUENZI, W. D. and BEARD, W. E. Anaerobic conversion of DDT to DDD and aerobic stability of DDT in soil, *Proceedings, Soil Science Society of America,* **32,** 522–524 (1968)

25. JENSEN, H. L. Notes on the microbiology of soil from Northern Greenland, *Meddelelser Grønland,* **142,** 23–29 (1951)

26. LYNCH, J. M. Identification of substrates and isolation of micro-organisms responsible for ethylene production in the soil, *Nature,* **240,** 45–46 (1972)

27. McCARTY, P. L. Energetics of organic matter degradation, in *Water Pollution Microbiology* (edited by R. Mitchell). Wiley/Interscience, New York (1972)

28. MEIKLEJOHN, J. Microbiology of the nitrogen cycle in some Ghana soils, *Empire Journal of Experimental Agriculture,* **30,** 115–126 (1962)

29. MEIKLEJOHN, J. Microbiology of Broadbalk soils, *Report of Rothamsted Experimental Station for 1968, Part 2,* 175–185 (1969)

30. MYERS, R. J. K. and McGARITY, J. W. Denitrification in undisturbed cores from a solodized solonetz B horizon, *Plant and Soil,* **37,** 81–89 (1972)

31. PAYNE, W. J. Reduction of nitrogenous oxides by micro-organisms, *Bacteriological Reviews,* **37,** 409–452 (1973)

32. PIRIE, N. W. (editor). *Leaf Protein: its Agronomy, Preparation, Quality and Use.* IBP Handbook No. 20. Blackwell, Oxford and Edinburgh (1971)

33. POSTGATE, J. R. and CAMPBELL, L. L. Classification of *Desulfovibrio* species, the nonsporulating sulfate-reducing bacteria, *Bacteriological Reviews,* **30,** 732–738 (1966)

34. *Report of Rothamsted Experimental Station for 1969, Part 1:* Nitrogen-fixing soil clostridia, 100–101 (1970)

35. *Report of Rothamsted Experimental Station for 1970, Part 1:* Nitrogen-fixing clostridia and Survey of *Clostridium botulinum* in soils, 91 (1971)

36. *Report of Rothamsted Experimental Station for 1971, Part 1:* Soil anaerobes and soil structure, 88–89 (1972)

37. RUSSELL, E. J. and APPLEYARD, A. The atmosphere of the soil: its composition and the causes of variation, *Journal of Agricultural Science* (Cambridge), **7,** 1–48 (1915)

38. RUSSELL, E. J. and RICHARDS, E. H. The changes taking place during the storage of farmyard manure, *Journal of Agricultural Science* (Cambridge), **8,** 495–563 (1917)

39. SKINNER, F. A. The enrichment and isolation of anaerobic cellulolytic soil bacteria, in *Anreicherungskultur und Mutantenauslese:* Symposium, Göttingen, 1964. *Zentralblatt für Bakteriologie, Parasitenkunde, Infektionskrankheiten und Hygiene,* Abt. I. Supplementheft 1, 91–94 (1965)

40. SKINNER, F. A. The anaerobic bacteria of soil, in *The Ecology of Soil Bacteria:*

Symposium, Liverpool, 1965 (edited by T. R. G. Gray and D. Parkinson). Liverpool University Press (1968)

41. SKINNER, F. A. The isolation of soil clostridia, in *Isolation of Anaerobes* (edited by D. A. Shapton and R. G. Board). The Society for Applied Bacteriology Technical Series No. 5. Academic Press, London (1971)

42. SMITH, K. A. and SCOTT RUSSELL, R. Occurrence of ethylene, and its significance, in anaerobic soil, *Nature,* **222,** 769–771 (1969)

43. WARING, S. A. and BREMNER, J. M. Ammonium production in soil under waterlogged conditions as an index of nitrogen availability, *Nature,* **201,** 951–952 (1964)

44. WILLIS, A. T. *Clostridia of Wound Infection.* Butterworths, London (1969)

45. WINOGRADSKY, S. '*Clostridium Pastorianum,* seine Morphologie, und seine Eigenschaften als Buttersäureferment', *Zentralblatt für Bakteriologie, Parasitenkunde, Infektionskrankheiten und Hygiene,* Abt. II, **9,** 43–54, 107–112 (1902)

46. WINOGRADSKY, S. Sur l'étude de l'anaérobiose dans la terre arable, *Compte Rendu Hébdomadiare des Séances de l'Académie des Sciences,* **179,** 861–863 (1924)

2 Rhizosphere Micro-organisms—Opportunists, Bandits or Benefactors

MARGARET E. BROWN

The large population of micro-organisms in soil lives in what is often described as an 'unstable equilibrium'—a state in which at any one time each individual is in balance with its neighbour but in which changes in environmental conditions lead to changes in the equilibrium. When soil is kept undisturbed and under constant conditions, the daily variability in the equilibrium is small, mainly because available energy sources are lacking. When plants are introduced into this environment, the whole situation for microbes changes drastically, because plants are among the chief suppliers of nutrients to the soil. It is not surprising, therefore, to find the microbial population exploiting this abundance, particularly in the root region, usually called the rhizosphere. The rhizosphere is not a uniform well-defined region but a zone with a microbial gradient extending from the root surface or rhizoplane, where micro-organisms are most affected, to soil a centimetre or two away, where effects are minimal, and the extent of this gradient depends principally on the plant species. Frequently the rhizosphere effect is expressed in terms of the R/S ratio—that is, the ratio of the number of micro-organisms in the rhizosphere soil to the number in the corresponding root-free soil. Roots strongly but selectively stimulate the multiplication of bacteria, fungi, actinomycetes and free-living nematodes. Algae and protozoa are not directly stimulated but increase in numbers because more bacteria means more food for them. (We owe much of our basic information about the rhizosphere to the Canadian school of microbiologists under the leadership first of A. G. Lochhead and then of H. Katznelson.)

As roots move through soil, root cap cells provide nutrients near the tip and elongating zone, but as plants age, root hairs and cortical cells die, the whole root eventually dies, and all this tissue becomes available to feed many microbes. Old and decaying roots support a large and varied population of species, not all of which are usually found in the more select microbial community of younger healthy roots; this suggests that cell debris is a less selective substrate than root exudates. The elongating portion of roots 1–3 cm from the apex is the major site of exudation, but older parts of roots also provide significant amounts of organic compounds and the lateral root zone, after full emergence, is also active, particularly at the tips of the laterals. Root hair tips also exude droplets. Certainly, intact healthy roots exude sufficient organic material to support large microbial populations—for example, roots of 2- and 8-week-old wheat grown in the field supported 1.5×10^6 bacteria per milligram of dry root.[3] With young wheat 1–2% of the carbon reaching the roots is released into soil: 0.2–0.4% as water-soluble exudate and 0.8–1.6% as insoluble mucilaginous material, including sloughed root cap cells. Dried wheat roots contain approximately 40% carbon; thus, the total organic material released may be 4–8 µg per milligram of dry root. Different experiments with plants of similar age, grown under identical culture conditions, show that the quantity and quality of compounds exuded differ considerably between species, each having a characteristic root microflora. The amount of nutrient determines the population size and the quality of the nutrient determines the nature of the associated microflora.

Root exudates contain small amounts of a miscellany of compounds, including sugars, amino acids, peptides, enzymes, vitamins, organic acids, nucleotides and, in trace amounts, various substances with specific biological activities such as nematode cyst-hatching factor and fungal zoospore attractants.[30] Some of these may diffuse away from the root or be adsorbed by the soil, but calculations of the expected microbial population based on the energy content of these materials agree well with actual counts made by orthodox dilution plate techniques, although these estimate only part of the root population.[3]

In most root exudate studies plants are grown under sterile conditions, frequently in nutrient solutions, obviously necessary if materials produced only by the plant are to be examined; otherwise contaminating micro-organisms will rapidly metabolise and change the amount and composition of the exudates, as well as altering the permeability of root cells or modifying root growth. Is it correct to assume that exudates from plants growing under conditions very different from those encountered in soil are the same as those produced in soil? Comparisons may be valid, but significance of results in relation to the rhizosphere effect should be assessed with caution.

An attempt to study exudates of tomato seedlings in soil was made by Subba Rao, Bidwell and Baily[35], using ^{14}C-labelled carbon dioxide and measuring the radioactivity which could be leached from the soil. Different tomato varieties exuded different amounts of radioactivity with a concentration gradient away from the roots. Only part of the radioactivity could be recovered after acid hydrolysis of the soil, which possibly represented exuded compounds fixed by micro-organisms.

Germinating seeds exude a rich supply of nutrients but the compounds are in different proportions compared with those from roots and usually feed a very mixed population. In a comparison of seed and seedling exudates from barley, wheat, cucumber and beans Vančura and Hanzlíková[39] found, in general, that the total nitrogen in seed exudates was constituted more from proteins and peptides than from free amino acids, whereas the reverse was true of root exudates. The amino acid spectrum was similar in both types of exudate, as were the organic acids, but the sugars present, especially keto sugars, differed considerably; also, more reducing sugars were exuded by roots than by seeds. The plants examined by Vančura and Hanzlíková[39] all had large seeds and their exudates showed differences in the relative amounts of each compound. The exudates from the infinite variety of plant species may provide a wide range and combination of substances, but the study of such exudates is very much in its infancy.

Environmental factors can influence the pattern of exudation: for example, high light intensity and temperature, temporary plant wilting and, of course, root damage all favour exudation. Differences in plant nutrition probably lead to changes in root exudates, although little work has been done on this important aspect. Foliar applications of certain nutrients, such as urea, cause marked increases in root exudation of glucose, fructose, glutamine and α-alanine and decreases in the amounts of organic acids.

Although selective for bacteria, root exudates seem to be non-selective for fungi, supporting saprophytes and pathogens alike and stimulating germination of many fungal spores. Infection of plants by soil-borne pathogens is an extensive subject and cannot be discussed in detail in this chapter, but interactions of temperature

23

and exudation on fungal infection can be cited as a good example of how the whole rhizosphere community reacts in a given situation. Root exudates of strawberries grown at 5 and 10 but not at 20 and 30°C stimulate spore germination and mycelial growth of *Rhizoctonia fragariae*—a pathogen causing much strawberry degeneration in cold soil but little decay in warm soil.[19] Low soil temperatures of 5–15°C tend to retard seed germination and seedling growth, but not active exudation, so that young plants remain for long periods in an environment particularly suited to the pathogen. With other plants and pathogens low temperatures may encourage microbes that compete with the pathogen for nutrients, thus reducing its infectivity, for many pathogens require a certain level of nutrition before they become infective. Alternatively, the microbes can produce antagonistic substances that inhibit the pathogen within the local environment of the seedling root. However, the whole subject of antibiosis as a disease-controlling mechanism is a vexed one and antibiosis may only rarely be involved. An outstanding example of disease prevention by the rhizosphere flora is given by the complete immunity of maize to root rot caused by *Phymatotrichum omnivorum* when maize is grown in soil or in non-sterile sand–bentonite mixture inoculated with the pathogen. Under sterile conditions—that is, without a rhizosphere microflora—the pathogen is lethal.[12]

Other factors contributing to the rhizosphere effect need much more study. The gaseous environment of the root is one example, but here there are many technical difficulties in obtaining measurements. The general composition of the gas phase in soil is scarcely relevant; it is the local concentrations of different gases at different positions along the root and on and within the soil crumbs which matter to the micro-organisms. We might expect that oxygen concentrations are low and carbon dioxide concentrations high because of root and microbe respiration, but recent work by Greenwood[18] on the concentration of carbon dioxide in the aqueous phase of aerobic soils has suggested that the growth of roots or micro-organisms is likely to be checked by O_2-deficiency before it is affected by an excess of carbon dioxide; thus the rhizosphere population may be one tolerating or even preferring conditions of lower O_2 tension for optimum growth.

Bacterial colonies develop in the zone of root elongation—the zone with most nutrient—within a few hours of root emergence. Initially they occur as widely separated, small clusters of cells, but as the root ages, the colonies increase in size and may eventually form a mantle over the root. Foster and Rovira[13] have studied the spatial distribution of micro-organisms on field-grown wheat roots, using ultra-thin sections, and found that colonisation of seminal roots is sparse and mainly at the junctions of cortical cells; young nodal roots are coated with a uniform layer of soil containing few bacteria, but old nodal roots have large aggregates of bacteria both inside and outside. Few bacteria on the seed coat contribute to the root population, although both groups of organisms can use seed and root exudates as substrate; presumably, those seed bacteria which do colonise roots have specific requirements which are met by the root environment. Recently an outer mucilaginous layer or 'mucigel' has been observed on roots and root hairs and especially around the root caps of plants grown either aseptically or in soil.[16] At least part of this 'mucigel' seems to be of plant origin, but microbes probably contribute to and also cause changes in its structure; electron micrographs indicate that 'mucigel' adjacent to micro-organisms is different from the bulk of the material. The 'mucigel' may be a substrate for micro-organisms, especially as it contains polysaccharides and pectic polymers (these have been identified), but its presence on non-sterile roots indicates either that its utilisation by microbes is limited or that perhaps the 'mucigel' is produced continuously along the root. Bacteria are embedded in the 'mucigel' and it may help pioneer species prevent subsequent microbial colonisation; also, it may influence movement of chemical compounds and micro-organisms between root and surrounding soil. As it may be a continuous layer, it may interfere with gaseous diffusion, thus helping to create anaerobiosis at the root surface. Perhaps its increased thickness at the root cap (some measurements give 2.5 μm) partly explains why micro-organisms rarely colonise this region, although the nutrient supply is plentiful with root cap cells and exudates. The 'mucigel' may preserve the rhizoplane population during drought conditions, as embedded organisms are less likely to be desiccated unless the root itself is so dried as to be killed. The rhizosphere population has no such protection and declines under dry conditions; what happens to the rhizoplane population under such conditions has not been determined. At present we can only speculate about the ecological and physiological significance of the 'mucigel'.

Differences between root and soil microflora can only be explained by the presence of different energy sources and, hence, of special environments, but neither the metabolic versatility of an organism alone nor any single nutritional or physiological character can account for the selection of certain species.

Given a selective stimulation of micro-organisms around roots, what is the role of this population? Is this microflora merely a population of opportunists taking advantage of the environment, giving nothing in return or even taking from the plant? Most answers to these questions relate to the role of bacteria, because these are the numerically dominant members of the population. Fungi increase numerically much less than bacteria, but this does not necessarily mean they are less important, because fewer units of large hyphae are required to be equivalent to thousands of bacteria. Little is known about the role of root fungi, neither those in the rhizosphere nor those firmly adhering to the root surface, but many experiments have produced the following basic facts about their growth.[27]

Fungal populations differ in the two zones of the root region and change as plants mature to a stable community of a few dominant genera. Few fungi from the seed colonise the root; instead root surface forms have their origins in the spores lying dormant in the soil until the nearness of the root stimulates germination. Fungi do not colonise roots until at least 24 h after germination, and probably successive lateral root colonisation is more important than longitudinal growth of mycelium down the roots from an initial point. During the middle period of plant life large areas of the root are free from fungi, but with maturity mycelial activity increases and the spaces and internal root tissues are colonised.

Many fungal propagules are subject to outside control in the form of soil fungistasis. Even when soil conditions are favourable, viable propagules do not germinate and mycelial growth is slowed or stopped, not because of endogenous dormancy, or temperature and moisture content of the soil, but because of other factors, which still remain undefined. Response to soil fungistasis is probably vital to the life cycle of any soil fungus, because most fungi have to survive periods when substrate is lacking or minimal, and if, during such a period, spores germinated spontaneously or hyphae grew, it would probably lead to starvation and death. Soil fungistasis controls such growth and is usually overcome by factors favouring development of the fungus. Fungistasis is overcome in the rhizosphere, spores germinate and mycelium grows around roots. This phenomenon has great economic importance, for induction and maintenance of and release from fungistasis are processes through which a soil-borne plant pathogen has to pass to maintain its inoculum potential between susceptible hosts. Any interference with these processes may lead either to decline or to increase of the pathogen.

Bacteria also seem to be subjected to a process akin to fungistasis, called bacteriostasis, which restricts growth, except when conditions are favourable. Bacteria selectively stimulated in the wheat rhizosphere do not multiply in soil away from the roots; the rhizosphere either lacks the bacteriostatic factor or supplies substances that are partially or completely absent in soil and essential for overcoming the inhibition.[5]

The bacteria in the root region are certainly opportunists, using the available nutrients for increased multiplication; generally, the stimulated bacteria are Gram-negative rods and pleomorphic forms belonging particularly to groups that ferment carbohydrates and decompose cellulose, ammonify and denitrify. These bacteria are more active physiologically, as shown by manometric experiments, when increased rates of oxygen uptake are obtained with substrates such as sucrose, glucose, acetate, succinate and alanine.[41] Most rhizosphere-to-soil ratios (*R/S*) range from 5 to 20, depending on plant species and age, and a larger proportion of rhizosphere than soil bacteria requires amino acids for maximum growth but is less fastidious over preformed organic compounds, vitamins and growth factors such as those found in yeast or soil extracts. The percentage of vitamin B_{12} requiring isolates from the rhizosphere is less than that from soil, but the actual number of these bacteria is greater, and many obtain their thiamin, biotin or vitamin B_{12} from the amino-acid-requiring bacteria which can synthesise these substances or from the fungi of the rhizoplane.[9] Abundance of various bacteria does not necessarily mean that any one group is more active than another: for example, although denitrifiers are stimulated, there is no evidence that, in general, denitrification is more rapid, but in local areas of anaerobiosis denitrification can be very important. It has been shown that 15–37% of fertiliser nitrogen was lost by volatilisation when applied to permanent grassland with water content below field capacity. Losses from nitrate fertiliser were double those from ammonium. This result was attributed to the large population of denitrifying bacteria, and to greater oxygen consumption and exudation of hydrogen donors by root systems.[31] An abundance of ammonifying bacteria means that more ammonia may be released from suitable substrates in the rhizosphere than in root-free soil.[33] Similarly, more rapid decomposition of amino acids involving ammonia release may be expected. Bacteria using carbohydrates or decomposing cellulose are plentiful around roots because of the substrate supply, but it is difficult to visualise how they may benefit the plant, although they probably help each other, particularly by breaking down cellulose to produce available carbon.

Nitrogen availability is affected by rhizosphere micro-organisms. Goring and Clark[15] showed that there was less nitrogen in the rhizosphere than the nitrate that would have been formed in unplanted soil. Nitrification was proceeding, but at the same time nitrogen was being immobilised by the micro-organisms.

Because of their need of minerals for growth, plants and rhizosphere microflora are always competing for them, and, where minerals are in short supply, the plant may be the first to suffer. The mantle of micro-organisms outside the root together with their short generation time means that they have first call on the minerals in the soil solution and rapidly assimilate what they need; this may create a deficiency for the plant. Alternatively, bacteria can adversely affect plants by immobilising essential elements. The deficiency diseases 'little leaf' of fruit trees and 'grey speck' of oats are caused by bacterial immobilisation of zinc and oxidation of manganese, respectively.[1]

An interesting example of the effect of the rhizosphere microflora on the molybdenum content of plants was described by Loutit and her co-workers.[22-24] In a dental survey in the Hawke's Bay area of New Zealand there were differences in the occurrence of dental caries in two districts—Napier and Hastings. Vegetables grown in Napier soils had higher concentrations of aluminium, molybdenum and titanium and lower concentrations of barium, copper, manganese and strontium than those grown in Hastings soil. Both soils were derived from a similar parent rock, but as a result of an earthquake a land mass that was previously a lagoon was raised 9 ft in the Napier district and was then used for market gardening and residences. The dental caries was thought to be related to the molybdenum content of the vegetables, and as the soils were similar in origin, it was suggested that the mineral uptake by the plants was affected by the microbial activity of the root zone. Radish was used as a test plant and it was found that numbers of micro-organisms in the rhizospheres of plants grown in the two soils were similar. However, plants growing in nutrient solution with an inoculum of Hastings rhizosphere organisms had less molybdenum than plants with a Napier inoculum. In cross-inoculation experiments with tomato and radish rhizosphere populations produced similar effects; the microflora from Hastings soil led to a lower concentration of molybdenum in the plants than the microflora from Napier soil.

It has also been shown that activity of rhizosphere micro-organisms leads to decreased uptake of other elements such as sulphur[34], calcium and rubidium.[38]

The effect of micro-organisms on uptake and translocation of phosphorus has attracted considerable interest and is discussed at length in another chapter. Twenty-five years ago Gerretsen[14] showed that uptake of phosphate by plants growing in sterilised sand containing insoluble phosphate compounds was increased by adding an inoculum of soil. Microbes isolated from the soil dissolved calcium phosphate suspended in nutrient agar. These findings prompted many investigations into the distribution of organisms capable of releasing phosphates from organic and inorganic phosphates, and such organisms have been found in many cultivated soils and are abundant in rhizospheres. However, results from phosphate uptake experiments done in sand culture cannot necessarily be extrapolated to soil, for it has recently been shown that soil added to sand culture or soil alone inhibits organic phosphate breakdown[17] and organisms able to mineralise organic phosphate in culture cannot do so in soil.[25] Nor is there positive evidence that soil micro-organisms are directly involved in liberating inorganic phosphate.

How, then, can the improved growth of Gerretsen's plants be explained? Are they larger because they took up more phosphate, or has some other factor caused increased growth which itself leads to increased phosphate uptake? Examination of the photographs

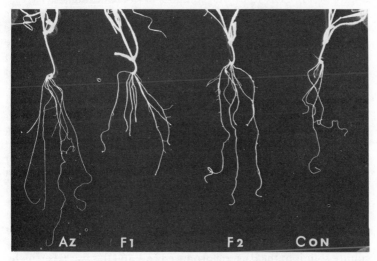

Figure 2.1 Effects of different soil micro-organisms on roots of wheat seedlings grown axenically.
Az = Azotobacter; F1, F2 = different species of fungi; Con = control

29

published in Gerretsen's paper shows that all the inoculated plants have elongated stem internodes and enlarged leaf areas and that flowering is advanced—all symptoms of root treatment with a small dose of growth-regulating hormone of the gibberellin type. It is now known that soil micro-organisms produce trace amounts of growth regulators which have marked effects on plant growth[4, 6] (see *Figure 2.1*), and the organisms used by Gerretsen could have produced such gibberellin-like substances.

The rhizosphere microflora, in general, probably affects plant growth by producing growth-regulating substances. Swaby[36] was one of the first people to show that plants grown in the presence of soil micro-organisms could be twice as tall as those grown under aseptic conditions. He suggested that microbes in association with organic matter frequently produced both stimulating and inhibiting substances; the two substances might be identical, causing stimulation of growth when present in the plant tissue in low concentrations, as in fast-growing plants, or depression when present in higher concentrations, as in slow-growing plants.

The root zone of wheat seedlings contains a significant proportion of bacteria that actively produce indolyl-3-acetic acid (IAA) when tryptophan is present in the test medium, but this population is not abundant in the root region of older plants.[4] Thus, young roots carry a population of bacteria able to produce IAA at a time when tryptophan is likely to be one component of root exudates. There is now ample evidence that roots can take up a variety of organic compounds, including indole compounds, and the region of uptake is most probably the zone of differentiation or where root hairs are most abundant—that is, the zone where the root microflora is also richest. Uptake of exogenous IAA thus increases the auxin content of the tissues.[32] Root growth is partly controlled by auxin[37]; decrease in the growth rate is apparently an indirect effect due to auxin-induced production of ethylene, which is the immediate inhibitor. Usually, inhibited roots are thickened and this results from interaction with cytokinin. Sometimes a brief exposure to auxin results in promotion of root growth; initially roots are inhibited but subsequently growth is promoted. At very low doses, about 175 µg/ml IAA, there is often growth promotion without inhibition. Brown[4] found that different rhizosphere bacteria produced from 0.02 to 0.3 µg/ml in cultures without L-tryptophan in the medium, and from 1.0 to 10.0 µg/ml with L-tryptophan. Libbert and Manteuffel[21] obtained significant increases in elongation and dry weight of maize growing in unsterile culture solution; the bacteria produced IAA and this was taken up by the plants.

30

Young wheat roots also carry a population of bacteria that inhibit growth of pea and lettuce plants when tested in bioassays, but this population decreases as plants age and is replaced by one producing growth-promoting substances of the gibberellin type; these bacteria are particularly abundant when the wheat is tillering and earing, a time of rapid growth expansion.[4] The amount of gibberellin-like substance produced per millilitre in different bacterial cultures ranged from 0.001 to 0.5 µg gibberellic acid GA_3 equivalents. Thus, roots in soil are exposed to a microbial population that may produce promoters and inhibitors.

It is possible to alter the rhizosphere microflora of young plants by introducing large numbers of micro-organisms, either by coating seeds with an inoculum or by dipping seedling roots into suspension of organisms before transplanting to the final growing position. The introduced organisms usually spread from the seed to the young roots but often, after several weeks, cannot be recovered from older roots; the inoculum cannot compete successfully with the natural rhizosphere flora. Sometimes the inoculum multiplies on young roots and then numbers remain stable until the plant is harvested.

When certain bacteria were used as inocula, it was possible to increase yields of crop plants, particularly of horticultural crops such as cabbage and lettuce—plants that are harvested in full leaf—or carrot, where the storage organ is the crop. Cereals sometimes respond to these inoculation treatments but increases in yield are usually less than 10%.[6]

The practice of this type of bacterial inoculation started in the Soviet Union in 1927, when mineral fertilisers were unavailable, and by 1958 about 10 million ha were treated. *Azotobacter chroococcum* was selected as an inoculant because it fixed atmospheric nitrogen and it was thought that nitrogen would be added to the soil; later a strain of *Bacillus megaterium* was also used because it was active in mineralising organic phosphorus compounds in culture media. Crop growth was improved, but certainly not because of nitrogen fixation and probably not because of phosphorus mineralisation. Other bacteria, principally species of *Pseudomonas, Clostridium* and *Bacillus* (other than *B. megaterium*), were also used, but less extensively, as inoculants. All the bacteria affected crop growth similarly, frequently increasing germination rates, elongating stems, enlarging leaves and causing earlier flowering and fruiting—effects indicative of a common cause, related not to mineralisation but to growth regulators (see *Figure 2.2*). The bacterial inoculants were more effective when used in conjunction with mineral and organic fertilisers and worked best in soils supporting good plant and

31

Figure 2.2 Growth of tomato plants after treating seedling roots at transplanting with a culture of Azotobacter

microbial growth. Plant response under these conditions is a good indication that phosphate mineralisation and increased uptake of phosphorus are not the causes of larger plants, for the amount of phosphate released by bacterial activity would be negligible in the presence of the available mineral fertiliser.

Tests by different people for production of growth regulators by the bacteria used as inoculants have shown that all produce traces of IAA and gibberellin-like substances, sufficient to cause changes in plant morphology; for only trace amounts, if applied at the right stage of plant development, will change growth completely. Brown, Jackson and Burlingham[8] found that one dose of 0.05 µg GA_3 applied at transplanting time to roots of tomato seedlings with expanding cotyledons (the time when primordia are differentiating) significantly changed plant growth through to development of the first flowering truss. The effective dose could have been less than 0.05 µg GA_3 because some might have been adsorbed on to soil particles and some metabolised before total absorption by the roots had occurred. *Azotobacter chroococcum* produces on average 0.05 µg GA_3 equivalents per millilitre of culture in 14 days; and when tomato seedling roots are treated with a culture, growth is changed drastically and similarly to treatment with authentic GA_3.[7] Bacterial inoculants, when applied to plants, probably contain sufficient growth regulator to influence future plant development, but as they also multiply around seedling roots (see *Figure 2.3*), possibly more hormone is produced and absorbed by the plants at the critical differentiation stage.

32

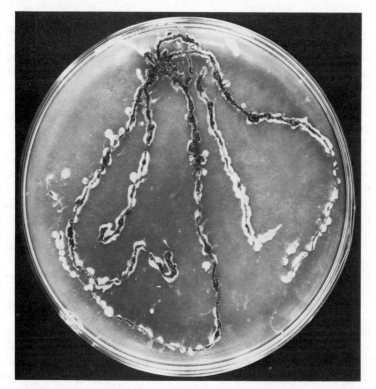

Figure 2.3 Development of Azotobacter *on roots of a wheat seedling, the grain of which was sprayed with* Azotobacter *culture before sowing in soil. The roots have been carefully removed from the soil and plated out on nitrogen-deficient agar*

The magnitude of the plant response will undoubtedly be influenced by the amount of hormone produced, which varies within and between species, and the environmental conditions, such as plant species, soil fertility and moisture, day length, light intensity, length of growing season and temperature; variations in any one of these conditions may explain why plant responses to inoculation are unpredictable and unreliable.

Recently it has been shown that extracts of several bacteria, including *Azotobacter chroococcum*[10] and *Rhizobium japonicum*,[28] possess a cytokinin type of activity. This growth regulator or related substances are implicated in the disease symptoms caused by *Agrobacterium tumefaciens*[29] and by *Corynebacterium fascians*,[20] and in inducing proliferation of root hairs in colza seedlings is induced when inoculated with *Arthrobacter* of rhizosphere origin.[2] Cytokinins are involved in all phases of plant development from cell division and enlargement to formation of flowers and fruits. They affect metabolism and the biosynthesis of growth factors, influence appearance of organelles and flow of assimilates and nutrients through the plant, and also enhance resistance to ageing and adverse environments. Obviously, if the rhizosphere microflora or inoculants produce these substances and the roots absorb them, plant growth will be affected.

It has also been suggested that bacterial inoculants act as suppressors of plant diseases, and an antibiotic isolated from cultures of *Azotobacter chroococcum* was found to suppress *Candida albicans, Alternaria* and *Monilia* species.[26] Suppression of diseases, especially those affecting seedlings, can improve yields of crops, particularly those sown from poor-quality seed, whose seedlings will be less vigorous and more prone to infection.

Although bacterial inoculation specifically for disease control has not proved very successful in natural soil, manipulating the rhizosphere microflora in more specialised environments has controlled some diseases. Inoculants which lyse, antagonise or compete with the pathogen may suppress it sufficiently to significantly decrease infection rate. Inoculation with *Bacillus subtilis* has successfully controlled seedling diseases in soils partially sterilised by aerated steam, and diseases of carnation cuttings grown in propagation media.[6] In some reports of successful disease control, the plants have also shown symptoms of treatment with a growth regulator; germination rate and root growth were improved and leaves and flowers stimulated.[6] This suggests that these substances may be implicated in disease control together with the direct effect of the inoculant on the pathogen. Treatment with IAA and GA$_3$ has altered disease expression in tomatoes infected with *Fusarium oxysporum* var. *lycopersici,* either by regulating the parasite growth and toxin production or by modifying the host response[11, 40]. Future research into biological control of diseases could profitably explore the possibilities of using inocula of mixed cultures of organisms all acting on the disease in different ways, such as inhibiting, competing for nutrients, providing growth regulators; used together with techniques for modifying the rhizosphere environment, we might find ways of controlling diseases that so far have not yielded to other controlling mechanisms.

Plants as we see them growing in soil are the products of their environment; the micro-organisms of the root region are part of that environment and by their activities for good or bad affect the product. As we learn more about the behaviour of these micro-organisms, we may be able to manipulate the population to one predominantly beneficial. This may not be easily achieved, for we are trying to change a population in which every member has its niche; manipulation may stimulate the activity of one or more groups of organisms and bring about a series of changes in soil with results difficult to predict. The aim of the soil microbiologist should be to favour the beneficial changes without the accompanying disasters which we have already learned can occur so easily when the balance of nature is upset.

REFERENCES

 1. BARBER, D. A. Micro-organisms and the inorganic nutrition of higher plants, *Annual Review of Plant Physiology,* **19,** 71–88 (1968)
 2. BLONDEAU, R. Production d'une substance de type cytokinine par des *Arthrobacter* d'origine rhizosphère, *Compte rendu (hebdomadaire) des séances de l'Academie des Sciences,* **270D,** 3158–3161 (1970)
 3. BOWEN, G. D. and ROVIRA, A. D. Are modelling approaches useful in rhizosphere biology?, in *Modern Methods in the Study of Microbial Ecology* (edited by T. Rosswell), 443–450. Swedish National Science Research Council, Stockholm (1973)
 4. BROWN, M. E. Plant growth substances produced by micro-organisms of soil rhizosphere, *Journal of Applied Bacteriology,* **35,** 443–451 (1972)
 5. BROWN, M. E. Soil bacteriostasis. Limitation in growth of soil and rhizosphere bacteria, *Canadian Journal of Microbiology,* **19,** 195–199 (1973)
 6. BROWN, M. E. Seed and root bacterization, *Annual Review of Phytopathology,* **12,** 311–331 (1974)
 7. BROWN, M. E. and BURLINGHAM, S. K. Production of plant growth substances by *Azotobacter chroococcum, Journal of General Microbiology,* **53,** 135–144 (1968)
 8. BROWN, M. E., JACKSON, R. M. and BURLINGHAM, S. K. Effects produced on tomato plants, *Lycopersicum esculentum,* by seed or root treatment with gibberellic acid and indol-3-yl-acetic acid, *Journal of Experimental Botany,* **19,** 544–552 (1968)
 9. COOK, F. D. and LOCHHEAD, A. G. Growth factor relationships of soil micro-organisms as affected by proximity to the plant root, *Canadian Journal of Microbiology,* **5,** 323–334 (1959)
10. COPPOLA, S., PERCUOCO, G., ZOINA, A. and PICCI, G. Citochinine in germi terricoli e relativo significato nei rapporti piante-microorganismi, *Annali di Microbiologia ed Enzimologia,* **21,** 45–53 (1971)
11. DAVIS, D. and DIMOND, A. E. Inducing disease resistance with plant growth regulators, *Phytopathology,* **43,** 137–140 (1953)
12. EATON, F. M. and RIGLER, N. E. Influence of carbohydrate levels and root surface microfloras on *Phymatotrichum* root rot in cotton and maize plants, *Journal of Agricultural Research,* **72,** 137–161 (1946)

13. FOSTER, R. C. and ROVIRA, A. D. The rhizosphere of wheat roots studied by electron microscopy of ultra thin sections, in *Modern Methods in the Study of Microbial Ecology* (edited by T. Rosswell), 93–95. Swedish National Science Research Council, Stockholm (1973)

14. GERRETSEN, F. C. The influence of micro-organisms on the phosphate intake by the plant, *Plant and Soil,* **1,** 51–81 (1948)

15. GORING, C. A. I. and CLARK, F. E. Influence of crop growth on mineralization of nitrogen in the soil, *Proceedings of the Soil Science Society of America,* **13,** 261–266 (1948)

16. GREAVES, M. P. and DARBYSHIRE, J. F. The ultrastructure of the mucilaginous layer on plant roots, *Soil Biology and Biochemistry,* **4,** 443–449 (1972)

17. GREAVES, M. P. and WEBLEY, D. M. The hydrolysis of *myo*-inositol hexaphosphate by soil micro-organisms, *Soil Biology and Biochemistry,* **1,** 37–43 (1969)

18. GREENWOOD, D. J. Distribution of carbon dioxide in the aqueous phase of aerobic soils, *Journal of Soil Science,* **21,** 314–329 (1970)

19. HUSSAIN, S. W. and McKEEN, W. E. Interactions between strawberry roots and *Rhizoctonia fragariae, Phytopathology,* **53,** 541–545 (1963)

20. KLÄMBT, D., THIES, G. and SKOOG, F. Isolation of cytokinins from *Corynebacterium fascians, Proceedings of the National Academy of Sciences,* **56,** 52–59 (1966)

21. LIBBERT, E. and MANTEUFFEL, R. Interactions between plants and epiphytic bacteria regarding their auxin metabolism. VII. The influence of the epiphytic bacteria on the amount of diffusable auxin from corn coleoptiles, *Physiologia Plantarum,* **23,** 93–98 (1970)

22. LOUTIT, M. W. and BROOKS, R. R. Rhizosphere organisms and molybdenum concentrations in plants, *Soil Biology and Biochemistry,* **2,** 131–135 (1970)

23. LOUTIT, M. W., LOUTIT, J. S. and BROOKS, R. R. Differences in molybdenum uptake by micro-organisms from the rhizosphere of *Raphanus sativus* L. grown in two soils of similar origin, *Plant and Soil,* **27,** 335–345 (1967)

24. LOUTIT, M. W., MALTHUS, R. S. and LOUTIT, J. S. The effect of soil micro-organisms on the concentration of molybdenum in the radish (*Raphanus sativus* L.) variety 'White Icicle', *New Zealand Journal of Agricultruai Research,* **11,** 420–434 (1967)

25. MARTIN, J. K. The influence of rhizosphere microflora on the availability of ^{32}P-*myo*-inositol hexaphosphate phosphorus to wheat, *Soil Biology and Biochemistry,* **5,** 473–483 (1973)

26. MISHUSTIN, E. N. Action d'Azotobacter sur les végétaux supérieurs, *Annales de l'Institut Pasteur,* **III,** suppl. 3, 121–135 (1966)

27. PARKINSON, D. Soil micro-organisms and plant roots, in *Soil Biology* (edited by A. Burges and F. Raw), 449–478. Academic Press, London (1967)

28. PHILLIPS, D. A. and TORREY, J. G. Cytokinin production by *Rhizobium japonicum, Physiologia Plantarum,* **23,** 1057–1063 (1970)

29. ROMANOW, I., CHALVIGNAC, M. A. and POCHON, J. Recherches sur la production d'une substance cytokinique par *Agrobacterium tumefaciens* (Smith et Town) Conn, *Annales de L'Institut Pasteur,* **117,** 58–63 (1969)

30. ROVIRA, A. D. Plant root exudates, *Botanical Reviews,* **35,** 35–57 (1969)

31. ROVIRA, A. D. and McDOUGALL, B. M. Microbiological and biochemical aspects of the rhizosphere, in *Soil Biochemistry* (edited by A. D. McLaren and G. H. Peterson), 417–463. Edward Arnold, London (1967)

32. SCOTT, T. K. Auxins and roots, *Annual Review of Physiology,* **23,** 235–258 (1972)

33. STARKEY, R. L. Some influences of the development of higher plants upon the micro-organisms in the soil. III. Influence of the stage plant growth upon some activities of the organisms, *Soil Science,* **27,** 433–44 (1929)

Rhizosphere Micro-organisms—Opportunists, Bandits or Benefactors

34. SUBBA RAO, N. S., BIDWELL, R. G. S. and BAILEY, D. L. The effect of rhizosphere fungi on the uptake and metabolism of nutrients by tomato plants, *Canadian Journal of Botany*, **39**, 1759–1764 (1961)
35. SUBBA RAO, N. S., BIDWELL, R. G. S. and BAILEY, D. L. Studies of rhizosphere activity by the use of isotopically labelled carbon, *Canadian Journal of Botany*, **40**, 203–212 (1962)
36. SWABY, R. L. Stimulation of plant growth by organic matter, *Journal of the Australian Institute of Agricultural Science*, **8**, 156–163 (1942)
37. THIMANN, K. K. The natural plant hormones, in *Plant Physiology. VIB. Physiology of Development. The Hormones* (edited by F. C. Steward), 3–145. Academic Press, London (1972)
38. TROLLDENIER, G. and MARCKWORDT, U. Untersuchungen über den Einfluss der Bodenmikroorganismen auf die Rubidium und Calciumaufnahme in Nährlösung wachsender Pflanzen, *Archiv für Mikrobiologie*, **43**, 148–151 (1962)
39. VANČURA, V. and HANZLÍKOVÁ, A. Root exudates of plants. IV. Differences in chemical composition of seed and seedling exudates, *Plant and Soil*, **36**, 271–282 (1972)
40. VOLKEN, P. Quelques aspects des relations hôte-parasite en fonction de traitements à l'acide indol-acétique et à l'acide gibberellique, *Phytopathologische Zeitschrift*, **75**, 163–174 (1972)
41. ZAGALLO, A. C. and KATZNELSON, H. Metabolic activity of bacterial isolates from wheat rhizosphere and control soil, *Journal of Bacteriology*, **73**, 760–764 (1957)

Suggestions for further reading

BAKER, R. Mechanisms of biological control of soil-borne pathogens, *Annual Review of Phytopathology*, **6**. 263–294 (1968)
BAKER, K. F. and SNYDER, W. C. (editors). *Ecology of Soil-borne plant pathogens*. University of California Press, Berkeley, California (1965)
COOPER, R. Bacterial fertilizers in the Soviet Union, *Soils and Fertilizers*, **22**, 327–333 (1959)
CLARK, F. E. Soil micro-organisms and plant roots, *Advances in Agronomy*, **1**, 241–288 (1949)
GRUEN, H. E. Auxins and fungi, *Annual Review of Plant Physiology*, **10**, 405–440 (1959)
KATZNELSON, H., LOCHHEAD, A. G. and TIMONIN, M. I. Soil micro-organisms and the rhizosphere, *Botanical Reviews*, **14**, 543–587 (1948)
LOCHHEAD, A. G. Soil microbiology, *Annual Review of Microbiology*, **6**, 185–206 (1952)
LOCKWOOD, J. L. Soil fungistasis, *Annual Review of Phytopathology*, **2**, 341–362 (1964)
MACURA, J. and VANČURA, V. (editors). *Plant Microbes Relationships*. Czechoslovak Academy of Sciences, Prague (1965)
MISHUSTIN, E. N. and SHIL'NIKOVA, V. K. *Biological Fixation of Atmospheric Nitrogen*. Macmillan, London (1971)
PARKINSON, D. and WAID, J. S. (editors). *The Ecology of Soil Fungi*. Liverpool University Press, Liverpool (1960)
PARKINSON, D. and GRAY, T. R. G. (editors). *The Ecology of Soil Bacteria*. Liverpool University Press, Liverpool (1967)
ROVIRA, A. D. Interactions between plant roots and soil micro-organisms, *Annual Review of Microbiology*, **19**, 241–266 (1965)

SCHROTH, M. N. and HILDEBRAND, D. C. Influence of plant exudates on root infecting fungi, *Annual Review of Phytopathology*, **2**, 101–132 (1964)

STARKEY, R. L. Inter-relations between micro-organisms and plant roots in the rhizosphere, *Bacteriological Reviews*, **32**, 154–172 (1958)

WATSON, A. G. and FORD, E. J. Soil fungistasis—a reappraisal, *Annual Review of Phytopathology*, **10**, 327–348 (1972)

3 A Microbiologist's View of Root Anatomy

BARBARA MOSSE

Soil micro-organisms perform a range of functions important for plant growth. By their activities in the nitrogen cycle and in the breakdown of organic matter and minerals, they release plant nutrients. Nitrogen-fixing organisms can increase soil nitrogen and mycorrhizal fungi effectively augment the absorbing surface of roots.[56] Some micro-organisms also have direct morphogenetic effects on plants.[14, 19] The extent of microbial activity depends much on plant exudates and residues, but the physical environment of the root can also be important. Some aspects of root form and function and their interrelationship with microbial and soil factors are discussed in this chapter.

3.1 The anatomy of roots in soil

Ontogeny and structure

Roots consist of an apical meristem producing cells in two directions: root cap cells towards the apex, and, in the opposite direction, cells that become the main body of the root. These cells originate in special zones of the meristem (histogen theory) determining their subsequent development into central cylinder (vascular tissue and pith), cortex or epidermis. Root cap cells are continuously sloughed off and replaced from the meristem. The origin of root cap and epidermal cells differs in different plant species[34] and this may affect the nature and extent of the mucilage layer. Two different developmental patterns are illustrated in *Figure 3.1*. In tobacco the epidermis, which develops from the protoderm, originates from cells ontogenetically related to the root cap. In maize the epidermis has no separate origin but differentiates

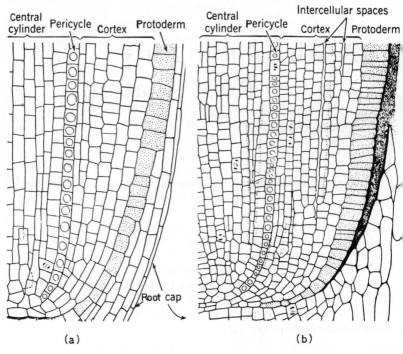

Figure 3.1 Drawing of longitudinal section of the root tip of (a) Nicotiana tabacum *and (b)* Zea mays *(after Esau[34])*

as the outer layer of the cortex. As the diagram shows, the epidermis in maize is covered by a mucilage layer from its inception. Four different patterns have been described in monocotyledons; in two, root cap cells arise from an independent meristem named the calyptrogen. A further complication arises from Clowes's demonstration of a 'quiescent centre' in the apical meristem of many roots.[21] Containing cells that divide rarely and synthesise DNA only slowly, this separates the meristem of the main root from that of the root cap. The central (columella) and outer root cap cells thus originate from distinct meristems, which makes the root cap heterogeneous in origin.

Behind the apical meristem is a region of elongating cells from which the central cylinder and cortex differentiate. The former comprises vascular elements, pericycle and endodermis. The endodermis is a compact layer of cells each with a suberised and possibly lignified band (the Casparian strip) along the radial and transverse walls. This impedes ion movement through the walls (free space) and ensures membrane-control of nutrient movement into and out of the vascular tissue. Cortical cells are usually large and rounded in cross-section, with extensive intercellular spaces between them. According to Scott,[66] cortical cell walls are lined with lipid material where they abut on intercellular spaces. In the Gramineae and some other species the spaces may be large (*Figure 3.1b*) and some result from cell lysogeny. This may account for the bacterial colonisation of some apparently dead cortical cells in some roots.[28] By contrast, cells of the central cylinder are close-fitting with few spaces.

The endodermis may give some protection against penetration by pathogens,[71] and neither ectotrophic nor vesicular–arbuscular (VA) endophytes ever penetrate it. The reasons are unknown. Garrett[40] considers that seedling roots, or young roots on older parts of the root system, are particularly susceptible to colonisation by un-specialised fungi because the juvenile tissues have not yet developed resistance to disease. The nature of this resistance is not discussed. In some plants calcium oxalate crystals form in the cortical cells; such cells are immune to infection by VA endophytes. Large applications of phosphate make most of the cortical cells in onion roots immune to VA infection. Only some sub-epidermal cells, which stain more deeply with Cotton Blue, remain suitable for intracellular growth of the endophyte.[57]

The longevity of the primary cortex varies in different species. The roots of most monocotyledons and some herbaceous dicotyledons are entirely primary. In others the primary cortex persists, although a cambium differentiates and some secondary growth is made. Where the primary cortex persists for a long time some suberisation and/or lignification may occur in the epidermal and sub-epidermal layers or in specialised cells. Roots of woody dicotyledons characteristically shed the primary cortex when secondary thickening takes place. A cork cambium is initiated in the pericycle, the primary cortex is sloughed off and the outer layer then consists of cork cells. In such plants the fibrous root system, which is renewed annually, is clearly distinguished from the persistent structural roots. Colonisation of the primary cortex generally increases with age[16, 64], but suberisation and particularly formation of an organised cork layer greatly impede root exudation and colonisation by micro-organisms. Ectotrophic mycorrhizal fungi increase the longevity of infected roots and delay their suberisation.[42] In roots where the primary cortex persists main and lateral roots are, nevertheless, often distinguished by diameter and numbers of cortical cell layers. VA endophytes also make some distinction in that they infect the smaller lateral roots more often and spread more widely in them than in the main roots. (*Table 3.1*).

Table 3.1

VA MYCORRHIZAL INFECTION IN MAIN AND LATERAL ROOTS
OF THREE PLANT SPECIES

Species	Main roots		Lateral roots	
	Incidence	Intensity	Incidence	Intensity
Lonicera periclymenum	33	7	42	76
Fragaria vesca	52	8	80	43
Mercuriales perennis	30	11	39	26

Incidence = % roots infected; intensity = % cortex colonised.

Root hairs

Nutrient uptake is one of the most important functions of the root system and root hairs, by virtue of their position on the root surface and their large surface-to-volume ratio, are well adapted to this function. Root hairs function for only a few weeks or even days, after which they collapse and become quickly colonised by micro-organisms. Sometimes root hairs become suberised or lignified, but when young, they are thin-walled and cytoplasmic movement may be seen in them. Often their surface is sticky and bacteria adhere to it. Most *Rhizobium* infections start in root hairs.

The root hair develops as an outgrowth from an epidermal cell which may be a specialised, shorter and less vacuolated cell (trichoblast) showing intense phosphatase and cytochrome activity.[5] The hair grows from a zone a few micrometres behind the root hair tip. Cellulose fibrils are laid down randomly at the tip but assume a longitudinal orientation as the hair elongates. Subsequent wall thickening near the root hair base consists of transverse or helically oriented fibrils.[22, 61] Currier[27] detected callose in *Hordeum* root hair tips by fluorescence microscopy. Often root hairs are covered by a mucilaginous layer containing pectin.[23, 24, 34, 66]

Cormack[23, 24] and others have found calcium and chelating agents to have marked morphogenetic effects on root hair development. He concluded that the root hair wall was two-layered, the more rigid outer layer consisting mainly of calcium pectate. Scott and co-workers[66, 67] proposed a more complex structure, resembling that of the leaf surface. They studied onion roots (which have no root hairs) and also other plants with root hairs, including *Vicia, Ricinus* and *Triticum*. They concluded that root hairs and the outer walls of epidermal cells have a fine cuticle-like layer containing lipids, covering a cell wall made of cellulose, other polysaccharides and pectic substances. The walls have pits with plasmodesma-like linings ending in the cuticle. These have been termed 'ectodesmata'. Through them solutions appeared to enter leaves and as many as 1000–1500 were recorded per 50 μm^2 of Primula leaf epidermis.[50] It is now doubtful whether any part of the protoplast extends into these channels.[38] Only Scott and her associates have reported them in roots. Those pictured in wheat root hairs[67] are less than 0.1 μm in diameter; such pits are too small for bacteria to enter but are interesting as possible pathways for exudation, and require further study.

While clearly desirable, a realistic assessment of amounts of viable and senescent root hairs is rarely possible in soil. One's ideas are often based on plants grown in agar, sand or solution culture or on the picture of a root emerging from a germinated seed on moist filter paper. Plant species is probably the most important single factor controlling root hair development. Figures of 5–17 μm root hair diameter and 80–1500 μm length were obtained from a study of 37 species in 20 families.[34] Most Gramineae have many long root hairs, while woody dicotyledons such as apple or citrus have short stubby ones. In two tropical legumes *Stylosanthes guyanensis (Figure 3.2)* and *Arachis hypogaea*[1] root hairs form only in axils of branch roots. Onions have none. Baylis[9] suggests that the mycotrophic habit might have evolved as an alternative to root hairs.

Environmental and nutritional factors affect root growth and some specifically affect root hair development. *Figure 3.3* shows a stylised picture published in 1883 of root hairs in water or moist air and in soil with decreasing water content. In sterile culture root hairs were more numerous, longer and developed closer to the apex than when soil micro-organisms were added.[14] Roots grown in sterilised soil in open pots generally are thinner, are more transparent, branch less and have more root hairs than those in soil re-inoculated with soil micro-organisms. Aeration can affect root hair development[70],

43

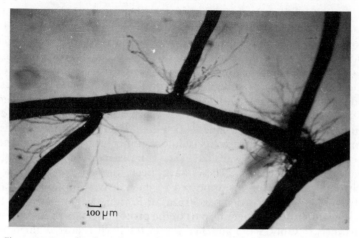

Figure 3.2 Root of Stylosanthes guyanensis *with root hairs in the axil of branch roots*

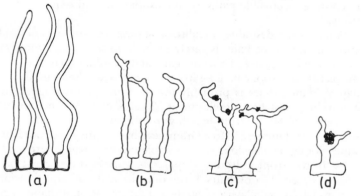

Figure 3.3 *Root hairs grown (a) in water or moist air, (b) in moist soil, (c and d) in dry soil (after Schwarz, 1883, quoted by Eames and McDaniel, 'An Introduction to Plant Anatomy', McGraw-Hill, 1925)*

and onions had root hairs when grown on filter paper moistened with culture solution.[62] In clover root organ cultures regions with few and many root hairs often alternate. Roots grown in soil and observed through glass usually have many root hairs in soil crevices and much fewer where they are in contact with soil. Root hair senescence often starts a few centimetres behind the root tip, but living nucleated hairs can occur on roots up to 15 cm long.[66]

44

The root cap and growth of roots through soil

The root cap is thought to protect the growing point as it pushes through the soil. Sloughed cells lubricate its passage and serve as substrates for micro-organisms. Nevertheless, it is difficult to visualise the process, to estimate how much damage is, done, and how much cell material sloughed off. Observing roots on damp filter paper, Rovira[63] thought that the root tip growing through soil would smear soil particles with non-diffusible substances (cells and polysaccharides), while the root behind the apex would exude small amounts of diffusible material. Exudation depends greatly on the amount of damage done. Contact with sand instead of culture solution increased it sevenfold,[17] and even careful handling[6] or contact with glass beads[8] increased it. Branch roots arising from the pericycle and pushing their way through the cortex also cause some dislocation and cell damage, leading to greater exudation, which may attract micro-organisms. For example, emerging laterals are a site for attachment of *Phytophthora parasitica* spores; *Thielaviopsis basicola* and *Fusarium solani* frequently enter tobacco or bean roots at points of lateral root emergence.[75] In wheat 1.2% of carbon reaching the roots was released into soil: 0.2–0.4% as water-soluble exudate, 0.8–1.6% as insoluble mucilage and sloughed root cap cells.[15] In plants where the entire primary cortex is sloughed this may well be the largest source of microbial substrate. Some may also be released from burst root hairs.

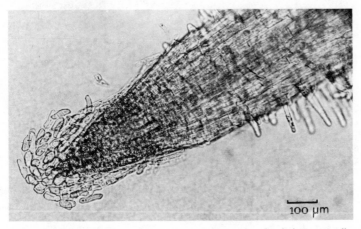

Figure 3.4 Root tip of clover grown on an agar medium. Note detached root cap cells

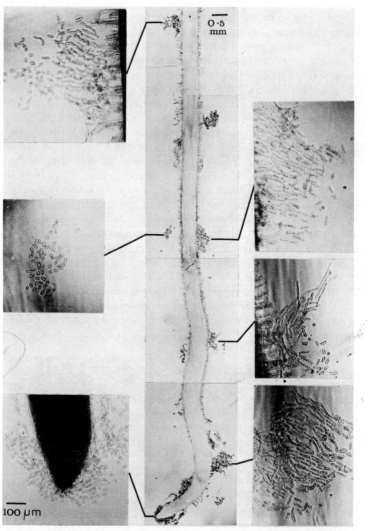

Figure 3.5 Clover root grown on an agar medium, showing the distribution and shape of sloughed cells

In maize, but not in other species, the root cap can be detached as a cohesive structure.[22] Root cap cells of clover grown on an agar medium, or in Fahraeus slides, appear to be largely separate (*Figure 3.4*), but when grown in liquid culture one finds some entire caps floating in the medium. Esau[34] also remarks on this effect of liquid culture. The degree of cohesion between the root cap cells, which might affect their ability to protect the root tip, may therefore depend partly on the environment.

In clover sloughed cells are of two kinds: small approximately isodiametric cells originating near the cap apex and long, often curled, narrow cells usually found where the cap joins the epidermis. In root organ cultures sloughed cells occur in small groups distributed at random along the root (*Figure 3.5*). What leads to their periodic release is not known. Main roots shed more cells than thinner shorter laterals. When shed, the root cap cells remain surrounded by a continuous wall layer. They have true plasmodesmata[22] which may act as channels of communication with the exterior when the cells are sloughed. If the sloughing of root cells in soil is similar to that in an agar medium, one should envisage such cells in small groups, attached to soil particles or the root surface, providing small localised pockets of substrate.

3.2 The root surface

Electron microscope studies have extended knowledge of the root surface in two respects: by giving more detailed pictures of the root surface at a magnification commensurate with the size of micro-organisms, and by providing information on the mucilaginous layer that often covers the root surface.

The shape of the root surface

It is rare to see a picture like that shown in *Figure 3.6,* but this may more accurately portray an oat root in its soil environment than the more usual textbook illustrations. Even a more normal-looking section of a young strawberry root (*Figure 3.7*) shows, on careful study, discontinuities and unevenness of the root surface with some damaged and collapsed epidermal cells. At electron microscope magnifications the surfaces of clover roots grown in agar (*Figure 3.8*) are often so irregular that it is difficult to decide what is outside and

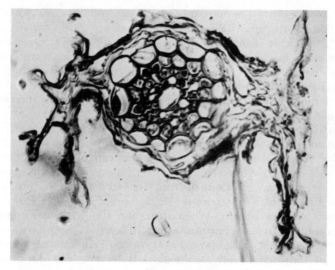

Figure 3.6. Transverse section of an old oat root grown in the field (after Strzemska, unpublished)

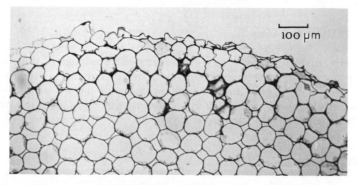

Figure 3.7 Transverse section of a field-grown strawberry root showing detail of the root surface
(by permission of D. Skene, East Malling Research Station)

what inside the root. Four-months-old seminal roots of wheat[37] and axenically grown pea roots[41] showed a similarly convoluted structure. Deep crevices are also evident between epidermal cells of barley roots grown in permutite.[43]

48

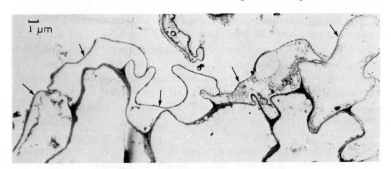

Figure 3.8 Electron micrograph of the surface of a clover root grown in an agar medium. External surface marked by arrows

Greaves and Darbyshire[41] comment on the damage in epidermal and outer cortical cells in the root hair zone and older parts of the root. Epidermal cells of *Ammophila arenaria,* a dune plant, were broken 4–5 mm and collapsed 40–45 mm from the root tip. Marchant[54] attributed this to mechanical damage by sand grains and also to microbial action. The extensive convolution of the root surface, the crevices and deeply embedded pockets open to the soil environment provide protected niches for micro-organisms and may explain the difficulty of sterilising roots by washing.

Root surfaces also have longitudinal corrugations, corresponding to the outline of the underlying cells. Scanning electron microscope pictures sometimes show these clearly.[20, 30, 54] Bacteria and fungi often spread along such furrows. This also occurs in leaves, where it is attributed to increased exudation at the cell junction, but in roots these sites may also offer protection against desiccation and abrasion.

The mucilage layer

Clowes and Juniper[22] define mucilage as a carbohydrate substance in plants, hard when dry, swelling and slimy when moist. Frey-Wyssling and Mühlethaler[39] distinguish two kinds of mucilage: cellulose mucilage is birefringent, reacts with zinc–chloride–iodine and contains microfibrils at the electron microscope level; while pectin mucilage is amorphous, stains with Ruthenium Red and Methyl Blue and is structureless.

49

Mucilage in plants has two origins. It may be secreted by living cells or it may arise by transformation of specialised cell walls. For example, epidermal walls of some seeds have a secondary apposition layer, which swells on germination and bursts the seed coat. The emerging seed is covered with mucilage.[39] In aquatic plants the epidermis secretes mucilage which can be regarded as analogous to the cutin layer of land plants.[34] Mucilage is commonly found associated with the root cap. It may form a layer, several millimetres thick, around air roots of tropical plants[34] and a substantial gelatinous sheath surrounds the root tip of many ericaceous plants.[52] In maize (*Figure 3.1*) and other plants a mucilage layer occurs between the root tip and the protoderm, and mucilage develops between cells of the root cap prior to sloughing (*Figure 3.4*). Young root hairs have a mucilaginous pectic covering.

Mucilaginous droplets form at the tip of maize roots. Their origin[44, 55] and chemical composition[35, 36] have been studied. The droplets originate from material contained in vesicles that form at the tip of cisternae of enlarged dictyosomes. The vesicles fuse with the plasmalemma and release their contents into the space between it and the cell wall. Under appropriate conditions this material is extruded through the wall and collects as a droplet at the root tip. Under the electron microscope it looks finely granular.

The droplets contain a highly hydrated polysaccharide, with an approximate molecular weight of 9×10^7.[36] On hydrolysis this yielded uronic acids, galactose, arabinose, xylose and fructose in the approximate molecular ratio of $3 : 7 : 5 : 11$. The droplets, which were not produced axenically, also contained free sugars (glucose and fructose) and had phosphatase and ATPase activity. Floyd and Ohlrogge[36] compared the absorption spectrum of the polysaccharide in the droplet with that of material from the mucilage thinly covering nodal roots grown in water-saturated sand. The mucilage was easily removed by washing in water. The two absorption spectra were similar but not identical. X-ray studies indicated that the exudate in the droplet might increase dispersal of the clay particles, exposing more active surfaces and thus affecting nutrient availability.[36]

Jenny and Grossenbacher's[43] electron micrographs of the root–'soil' boundary of barley roots grown in permutite greatly stimulated the interest of soil microbiologists in the mucilage layer. Their pictures showed an electron-transparent zone, sometimes containing bacteria, on the outside of the root. At the outer edge of this zone, not otherwise distinguishable, marker particles of ferric hydroxide accumulated. Where the mucilage touched a permutite particle it ceased to have a boundary of its own, behaving more like a liquid. 'Here', say the authors, 'the concept of a self-contained mucilaginous layer or film loses its proper meaning, and we prefer to speak of a mucilaginous gel or mucigel.' The term 'mucigel' was therefore intended to describe specifically a material or exudate in a semi-liquid state, which impregnated adjacent permutite or soil particles.

Several electron microscope and Stereoscan pictures have been published of the mucilage layer in soil and solution-grown plants.[18, 20, 29, 31, 33, 37, 41] In the most complete study Greaves and Darbyshire[41] examined roots of 16 agricultural plants grown in garden soil and in sterile soil, sand and solution culture. In sterile culture the mucilage appeared as an unevenly distributed layer of granular and fibrillar material covering the root surface and most root hairs. Most mucilage occurred round the root cap. Soil particles adhered to the outer surface of the mucilage layer, or were distributed within it. In some pictures soil particles also adhere to root hair surfaces without a visible mucilage layer.[29] Particularly older roots grown in unsterile soil and roots in sterile soil or solutions inoculated with micro-organisms have many bacteria and fungi embedded in the mucilage layer.[31, 37, 41] The bacteria are often surrounded by an electron-transparent zone, probably their own polysaccharide, which is distinct from the surrounding mucilage. When axenic cultures were inoculated with bacteria, larger quantities of mucilage developed, there were more dead or damaged root surface cells and the mucilage layer often had a distinct outer boundary.[29] This membrane-like structure often appears as a single dark line; sometimes it has a more complex structure. Occasionally it appears continuous, with an outer layer (perhaps the cuticle reported by Scott[66]) covering the epidermis.

Many problems concerning the origin, physical state and function of the mucilage layer remain unresolved. The mucilage in the root cap evidently has two separate origins, from mucilaginised cell walls and from active secretion. Possibly the two kinds of mucilage structure, granular and fibrillar, reflect this different origin.

It is not known whether mucilage droplets are secreted from root hairs and epidermal cells as well as from the root tip, or how common the phenomenon is in different plant species. It has been observed in maize and apple. Mature root hairs and epidermal cells contain little cytoplasm and few dictyosomes and secretion, if by the same mechanism as in the root cap, would not continue for long. If most of the mucilage is a semi-liquid secretion from the root cap, it

51

would probably impregnate soil particles touching the root tip. These may subsequently adhere to the root surface and be the soil crumbs used to count rhizosphere populations.

The membrane-like boundary of the mucilage layer also raises questions, particularly as it seems to occur also in solution-grown plants.[31, 41] If it is of biological origin, it might be the epidermal cuticle, in which case the mucilage would probably be a secretion from the root rather than the root cap. If so, how do bacteria get into the mucilage layer without damaging the membrane? It might also be due to the presence of micro-organisms on or within the mucilage, a possibility considered by Greaves and Darbyshire.[41] Perhaps the more likely explanation is a non-biological origin of the boundary membrane. It might form by surface drying or like skin on custard by chemical interaction with the environment. The dense 'mucilage' layer on *Paspalum notatum* roots[33] is a hard, reddish-brown, firmly adhering layer, only removable by vigorous scraping. It appears to be heavily impregnated with iron from the soil.

It is widely assumed that the mucilage layer provides a rich substrate for micro-organisms. This may not always be so, particularly when it consists largely of degraded wall material. One rarely sees any sign of change or degradation in the mucilage surrounding bacteria, and the large accumulation of their own polysaccharide suggests encapsulation rather than vigorous growth.

A recent observation on clover root organ cultures growing in an agar medium[58] may provide a technique for further study of the mucigel. Placing a drop of Toluidine Blue on roots growing on the agar surface produces immediate coagulation and a bright crimson colour in a material which seems to exist as a semi-liquid film over the root, and particularly the root cap (*Figure 3.9*). The stained film

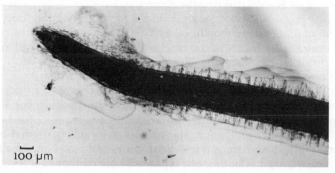

100 µm

Figure 3.9 Clover root grown on an agar medium and stained with Toluidine Blue to show the mucigel

can, with care, be lifted off the root and agar surfaces. At its outer edge it has a clearly defined boundary, but next to the root it is more diffuse and separates cleanly from it. The crimson colour indicates an alkaline reaction.

There is some evidence that the function of mucilage in plants is connected with its ability to swell or to harden. Primarily a protective and lubricating layer around roots, it may also provide anchorage for micro-organisms near the root surface, and possibly acts as a diffusion barrier for root exudates. Much clearly depends on its physical state, which probably varies with soil moisture.

3.3 Effects of micro-organisms on root anatomy

Local lesions, hypertrophy and hyperplasia, and production of gums, tannins and phenolic compounds are frequent plant reactions to invasion by pathogenic organisms. Less extreme reactions are reduced root growth, fewer and shorter root hairs and more branching. Symbiotic associations between roots and micro-organisms sometimes cause striking macroscopic changes in root structure with formation of a new dual organ. Such are the nodules of legumes and the ectotrophic mycorrhiza of some trees.

Nodules

The 'nodules' that form on roots last from a few months to several years and grow from a few millimetres to several centimetres. They vary much in shape and are born singly or in clusters, on tap roots, in the axils of branch roots or on ordinary roots.[12, 32, 48, 51] They differ in their anatomy, in the micro-organisms that induce and/or inhabit them, and in their ability to fix atmospheric nitrogen. Many of these structures are modified lateral roots, while others develop adventitiously in the cortex. None has a root cap.

In the Podocarpaceae, Araucariaceae and Sciadopityaceae the fine roots are studded with small, uniform nodules aligned in longitudinal rows (*Figure 3.10*). Usually less than 1 mm,[10] these nodules appear to form spontaneously without stimulation by any micro-organism; Khan[46] produced such nodules axenically in *Podocarpus falcatus*. They are modified lateral roots, arising, like them, in the pericycle of the parent root and pushing through the overlying cortex. They then cease to elongate and when mature have a central vascular cylinder that extends into the nodule. Mature nodules have

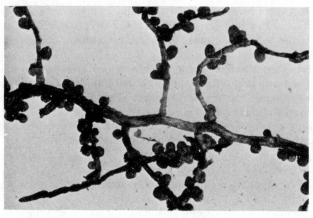

Figure 3.10 Nodules on Podocarpus dacrydioides *roots (by permission of G. T. S. Baylis)*

no apical meristem. They may have root hairs, but in Autumn the nodule cortex lignifies and the cells die. The nodules are nevertheless perennial; in Spring meristematic activity resumes in the pericycle at the tip of the vascular strand and a new nodule emerges from the top of the old.[10] The parent roots may have a beaded appearance. The cortex of such nodules is often colonised by VA endophytes, species of *Endogone,* forming typical VA infections. Electron microscope pictures of the infection in *Podocarp* nodules[11] look very similar to that in roots of *Ornithogalum,*[65] tobacco[45] and onion.[25] Nodules of *Podocarpus dacrydioides* grown in perlite also contained a mixed population of bacteria, chance contaminants amounting to 5×10^6 organisms per gram of root.[10] Two other species—*Aesculus indica*[47] and *Tribulus cistoides,*[2] a tropical herb in the Zygophyllaceae—form similar nodules spontaneously. Those of *Aesculus* are also inhabited by VA endophytes.[47]

Nodules induced by invasion of micro-organisms are of two kinds: those occurring in legumes and caused by species of *Rhizobium;* and those caused by an obligate symbiont, thought to be an Actinomycete, in a range of plants, often woody, in the Casuarinaceae, Myricaceae, Betulaceae, Eleagnaceae, Rhamnaceae, Coriariaceae and Rosaceae.[12] Both types fix atmospheric nitrogen, but they differ in nodule structure. Like the Podocarp nodules, the non-leguminous nodules are modified lateral roots originating within the pericycle of the parent root. Where infection could be observed the endophyte entered the curled root hair tip and spread into the cortex. A branch root arising in the pericycle near the infected cortical cells then gave rise to the nodule.[12, 13] The mature nodule has a central vascular cylinder enclosed by an endodermis often containing tannin, an enlarged cortex and a few layers of cork cells.[13] The apical meristem, except in Casuarinaceae and Myricaceae, branches dichotomously (*Figure 3.11*), so that the nodules have a corralloid appearance. In *Casuarina, Gale* and *Myrica* a normal, uninfected, but negatively geotropic root grows from each nodule lobe. Not all cells of the cortex are infected; their distribution varies according to the host species.[13]

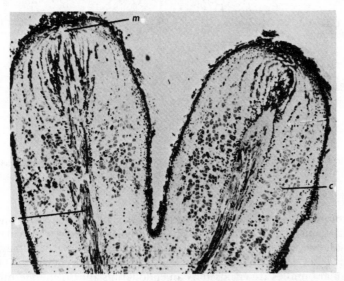

Figure 3.11 Section through part of a branched nodule of Alnus glutinosa: *m = meristem, s = vascular tissue, c = cortex with endophyte-containing cells (after Bond[13])*

The initiation and development of leguminous nodules have been extensively studied and described. Only the origin and structure of the mature nodule are relevant here. In most temperate legumes bacterial entry is through root hairs. It is preceded by a characteristic curling of the root hair, thought to be induced by hormones secreted by the bacteria or induced in the plant. The curling would affect the orientation of cellulose microfibrils and probably also their rate of deposition. Occasionally infection threads develop without previous curling where two root hairs are touching. The physical environment

55

of an enclosed space, existing also within the crook of a curled hair, may be as important for entry as physiological changes in the curled hair. Bacteria penetrate into the cortex by means of an infection thread and its vicinity stimulates some cells of the inner cortex to divide and initiate a meristem. This becomes the apical meristem from which the entire nodule develops. *Figure 3.12* illustrates the

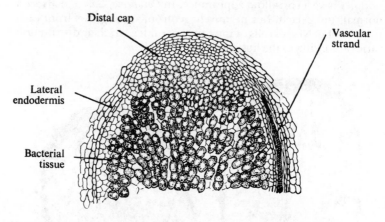

Distal cap

Vascular strand

Lateral endodermis

Bacterial tissue

Figure 3.12 Section of upper half of a nodule of Medicago sativa L. *(after Thornton and Rudorf, Proc. Roy. Soc. B,* **120**, *240 (1936)*

anatomy of a lucerne nodule. Characteristic is the large central zone which contains the bacteria. It is surrounded by discrete vascular bundles developing basipetally, from the nodule tip towards its base, where they join with the vascular system of the parent root. The vascular bundles are enclosed by an endodermis and an endodermis also separates the bacterial central tissue from the cortex that surrounds it. Two features (*Figure 3.13*) distinguish this nodule from all those previously described: (1) The zone colonised by micro-organisms lies in the centre of the nodule and the cortex is free from infection. In all the other nodules the vascular tissue lies in the centre of the nodule and the cortex is colonised. (2) The nodule arises from the cortex, exogenously, and later establishes vascular connection with the parent root basipetally. Other nodules arise endogenously from the pericycle and the vascular tissue develops acropetally; i.e. it grows from the parent root with the expanding nodule. The histological nature of the bacterised central zone of leguminous nodules is not clear. It could represent a pith. Except for the absence

Rhizobium nodule   Other nodules

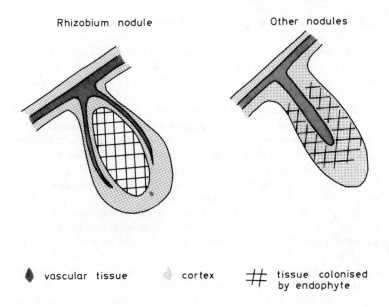

♦ vascular tissue   cortex   ╫ tissue colonised
                                      by endophyte

Figure 3.13 Diagram illustrating the difference between nodules formed by rhizobia and those formed or inhabited by other endophytes

of bud and leaf initials, *Figure 3.12* resembles a shoot apex. It is perhaps of some interest that nodule-like structures called sphaeroblasts can also form exogenously in the cortex of some stems.[7] Either they become lignified or, if the underlying cambium is stimulated to produce callus tissue and establishes connection with the still meristematic nodule, a shoot may be formed. This has the attribute of juvenility, i.e. it is in the physiological state of a seedling, and has the same leaf shape and capacity to form adventitious roots, which is often lost in the mature form. Recently Trinick[73] showed that nitrogen-fixing nodules of a non-leguminous tree, *Trema aspera,* belonging to the Ulmaceae, were inhabited by rhizobia able to form typical nodules with legumes. Interestingly, the *Trema* nodules conform to the usual legume pattern with a central bacterised zone, peripheral vascular system and cortical parenchyma.

Slight unusual for a legume is the development of nodules in the peanut, *Arachis hypogaea,* although the structure of the mature nodule follows the usual legume pattern with a bacteroidal central

57

area enclosed by an uninfected cortex.[1] Bacterial entry occurs through ruptured tissue at the site of lateral root emergence. The nodule develops in the tissue of the emerging lateral root and is enclosed at that time by the endodermis of the parent root. Allen and Allen[1] therefore considered it to be endogenous. However, separate tissues can sometimes be distinguished in the emerging lateral root while it is still enclosed by the stretched endodermis of the parent root,[34] and the pictures of the peanut nodules[1] would support the view that they arise from cortical initials of the emerging lateral root.

A range of nodule-like outgrowths have been produced in various plant roots by treatment with growth substances.[3, 4, 60] Many produce lateral roots or unorganised tissue. Most like legume nodules are the outgrowths produced by cytokinins in *Alnus*[60] and in tobacco root cultures.[4] In *Alnus* meristematic centres arose in the cortex and gave rise to whorls of tissue sometimes containing tracheids. These growths again resemble sphaeroblasts. In tobacco pseudo-nodules developed over the entire root system. At maturity they were 3–5 mm long, rough-surfaced, spherical or elongated. The nodules originated from groups of actively dividing enlarged cortical cells. Sometimes branch roots arose below these and then nodule-like outgrowths developed in the axil of the branch root, as in *Arachis*. However, the 'nodular' tissue neither contained defined meristems nor formed vascular connection with the parent root. Apparently no attempt was made to inoculate these nodules with rhizobia. In view of occasionally expressed hopes to induce nitrogen-fixing nodules in non-leguminous crop plants, the induction of exogenously produced, nodule-like structures by hormone treatment is interesting.

MacGregor and Alexander[53] produced tumour-like outgrowths on some legume roots by inoculation with a non-nodulating strain of *Rhizobium* and with irradiated cultures of two nodulating strains. *Agrobacterium tumefaciens* induced similar tumours in some legumes and also caused the cortex to split.

Nodules, then, can be formed spontaneously or induced by micro-organisms. They are of two main kinds: arrested lateral roots or new formations from cortical tissue. Both kinds can fix nitrogen but the organisms involved differ. In addition, arrested lateral root nodules can be colonised by vesicular–arbuscular endophytes. Macroscopically all these nodules can be quite similar. The kind of nodule formed is usually, but not invariably, determined by botanical affinity of the host.

Mycorrhiza

In general, only ectotrophic (sheathing) mycorrhiza exhibit marked morphological and anatomical modifications in root structure. In orchids and ericaceous plants root morphology is little changed by mycorrhizal infection. Infections by vesicular–arbuscular endophytes may cause a slight thickening of the cortex, increased branching of fine roots and a yellowish-green coloration. The colour is due to a water-soluble substance formed chiefly in cells with arbuscules. In some tree species,[48, 74] notably *Acer*,[49] intermittent root growth, which can be caused by drought, leads to a beaded appearance of the absorbing rootlets. In such roots cortical cells are extended radially rather than longitudinally, the root tips become suberised and the endodermal cells become impregnated with a tannin-like substance. Under favourable conditions growth is resumed from the root tip. Such beaded roots are frequently colonised by VA endophytes, though the endophyte does not induce the beading and also infects normal roots of the same plant.

Ectotrophic mycorrhiza comprise a very varied group of structures and associations (*Figure 3.14*). The morphogenesis of beech and pine mycorrhiza has been studied in detail and the findings have been

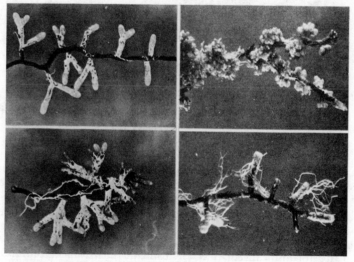

Figure 3.14 Different ectotrophic mycorrhizal roots of Pinus contorta *(after Zak[76])*

reviewed by Harley.[42] The extensive literature on the appearance and classification of ectotrophic mycorrhiza in general has recently been summarised by Zak.[76] As a single host species (*Pseudotsuga menziesii*) may bear over 100 different ectomycorrhiza,[76] it is evident that only the main modifications of root structure resulting from such associations can be considered here.

The typical ectotrophic mycorrhiza differs markedly from the uninfected root (*Figure 3.15*). It is a short lateral rootlet that has

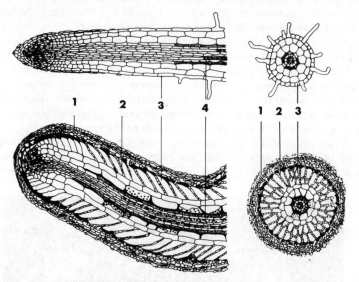

Figure 3.15 Longitudinal and transverse sections (diagram) of non-mycorrhizal (above) and mycorrhizal (below) Eucalypt roots (after Chilvers and Pryor, Australian J. Botany, **13**, 245 (1965)). 1 = fungal mantle, 2 = Hartig net, 3 = epidermal cell, 4 = vascular tissue

ceased growing and is fully differentiated up to its apex. The outer surface, which can be smooth, hairy or even spiny, is usually covered by a pseudoparenchymatous tissue made up of closely adpressed hyphae. This mantle or sheath envelops the entire rootlet, including the apex. It originates as a fine weft or thin discontinuous sheet of hyphae on the root surface and is the first stage in the development of an ectotrophic mycorrhiza. Occasionally the mantle may be absent or consist of a single layer of hyphae, but usually it is several microns thick. Sometimes it consists of two layers. The inner layer may contain dead, sloughed root cells, but there are no root hairs and no root cap comparable to that of the uninfected root. The outer cells of the root cortex are greatly modified in structure. They extend radially rather than longitudinally, may contain tannin and often have much thickened walls. These outer and some underlying cell layers may be colonised by the fungus, either intercellularly or intracellularly. The intracellular infection consists of masses of interwoven hyphae somewhat similar to those in ericaceous and orchid mycorrhiza but with little evidence of fungal digestion. Such infections are sometimes described as ectendotrophic. More typical is an intercellular infection—'the Hartig net'—consisting of septate hyphae that grow between the cortical cells. Usually these only penetrate the more superficial cortical layers. The endodermis is never penetrated and the structure of the stele remains unaltered.

The typical anatomy and morphology of ectotrophic mycorrhiza can be simulated under axenic conditions by auxin treatments and by culture filtrates from ectotrophic mycorrhizal fungi.[42, 69] Slankis[68] showed that auxins, applied exogenously, could be translocated both distally and proximally within the root. He concluded that auxin produced by mycorrhizal fungi caused the formation of short roots, a prerequisite for mycorrhiza formation, and influenced the whole root system, determining the frequency and sequence of long and short root production. This perhaps explains why uninfected short roots, occasionally infected long roots and other intermediate stages between uninfected roots and typical ectotrophic mycorrhiza occur in mature trees in natural habitats.[42] Slankis[69] also showed that at low light intensity the morphogenetic effects of auxin could be suppressed.

Macroscopically, ectotrophic mycorrhiza can be single swollen laterals, complex dichotomously branched, corralloid, nodular or tuberculate structures. The tuberculate[72] and nodular[42] structures consist of a peridium or a loose weft of hyphae enclosing a much-branched mycorrhizal rootlet; sometimes fungal fructifications attached loosely to the root have been described as tuberculate mycorrhiza.

Dual infections

A single host plant can react to different micro-organisms, providing for each its required habitat. Thus, different infections can occur close together on the same root system but they are never mixed. For instance, VA mycorrhiza can infect roots of legumes but will not invade rhizobial nodules borne on the mycorrhizal root.[26] Similarly, in *Discaria* actinomycete-containing nodules can arise from roots infected with VA endophytes. Whether the fungal endophyte infects the roots before, after or during nodule formation is not known. *Alnus rubra* has several kinds of ectotrophic mycorrhiza[59] as well as actinomycete-containing nodules, and *Dryas* species also have both. Ectotrophic and VA mycorrhiza can occur on different parts of the same hazel root. Only *Rhizobium* and actinomycete-containing nodules appear, so far, restricted to separate hosts.

3.4 Conclusion

Soil microbiology is concerned with the classification, enumeration and activities of micro-organisms in soil, and many microbial activities affect plant growth, directly or indirectly. Direct effects largely depend on the proximity of micro-organisms to the root. If bacteria increase the availability of nutrients, such as phosphate, secrete growth substances or fix nitrogen at a distance from the root, the effect of these activities will be diminished in proportion to the distance of the micro-organisms from the root. Spatial relationships are therefore important and I have tried to give some account of these.

A reciprocal relationship exists between microbial and root activities which finds its greatest expression where modifications in root structure create specialised habitats for micro-organisms. Such modifications can sometimes be simulated by treating axenically grown plants with hormones. It is reasonable to assume that this is also the mechanism whereby micro-organisms induce such structural changes. Very recently it was shown that the profusely branched roots characteristic of the Proteaceae can be induced by micro-organisms but are not inhabited by them. *Podocarpus falcatus,* on the other hand, produces nodules axenically and other Podocarpaceae are also thought to produce nodules spontaneously. Subsequently these may be colonised by endophytes. One wonders whether there might be some common evolutionary trend in these associations or whether each has developed along its own particular pathway.

Anatomical and morphological studies stand in some danger of being superseded by electron microscopy and the new insight it gives into host–micro-organism relationships. The two techniques are complementary, and it would be a pity if the knowledge gained from the earlier studies of developmental anatomy of these associations should be forgotten or the value of this approach overlooked.

REFERENCES

1. ALLEN, O. N. and ALLEN, E. K. Response of the peanut plant to inoculation with rhizobia, with special reference to morphological development of the nodules, *Botanical Gazette*, **102**, 121–142 (1940)
2. ALLEN, E K. and ALLEN, O. N. The anatomy of the nodular growths on the roots of *Tribulus cistoides* L., *Soil Science Society of America Proceedings*, **14**, 179–183 (1949)
3. ALLEN, E. K., ALLEN, O. N. and NEWMAN, A. S. Pseudo-nodulation of leguminous plants induced by 2-bromo-3,5-dichlorobenzoic acid, *American Journal of Botany*, **40**, 429–435 (1953)
4. ARORA, N., SKOOG, F. and ALLEN, O. N. Kinetin induced pseudonodules on tobacco roots, *American Journal of Botany*, **46**, 610–613 (1959)
5. AVERS, C. J. Histochemical localisation of enzyme activity in the root epidermis of *Phleum pratense*, *American Journal of Botany*, **45**, 609–613 (1958)
6. AYERS, W. A. and THORNTON, R. H. Exudation of amino acids by intact and damaged roots of wheat and peas, *Plant and Soil*, **20**, 364–370 (1968)
7. BALDINI, E. and MOSSE, B. Observations on the origin and development of sphaeroblasts in the apple, *Journal of Horticultural Science*, **31**, 156–162 (1956)
8. BARBER, D. A. and LEE, R. B. The effect of microorganisms on the absorption of manganese by plants, *New Phytologist*, **73**, 97–106 (1974)
9. BAYLIS, G. T. S. Fungi, phosphorus and the evolution of root systems, *Search*, **3**, 257–258 (1972)
10. BAYLIS, G. T. S., McNAB, R. F. R. and MORRISON, T. M. The mycorrhizal nodules of Podocarps, *Transactions of the British Mycological Society*, **46**, 378–384 (1963)
11. BECKING, J. H. Nitrogen fixation and mycorrhiza in Podocarpus root nodules, *Plant and Soil*, **23**, 213–226 (1965)
12. BECKING, J. H. Plant-endophyte symbiosis in non-leguminous plants, *Plant and Soil*, **32**, 611–654 (1970)
13. BOND, G. The root nodules of non-leguminous angiosperms, in *Symbiotic Associations:* XIIIth Symposium of the Society of General Microbiology (edited by P. S. Nutman and B. Mosse), 72–91 (1963)
14. BOWEN, G. D. and ROVIRA, A. D. The influence of micro-organisms on growth and metabolism of plant roots in *Root Growth:* Proceedings of 15th Easter School in Agricultural Science University of Nottingham (edited by W. J. Whittington), 170–199. Butterworths, London (1969)
15. BOWEN, G. D. and ROVIRA, A. D. Are modelling approaches useful in rhizosphere biology?, *Bulletins from the Ecological Research Committee* (Stockholm), **17**, 443–450 (1973)
16. BOWEN, G. D. and THEODOROU, C. Growth of ectomycorrhizal fungi around seeds and roots, in *Ectomycorrhizae* (edited by G. C. Marks and T. T. Kozlowski), 107–150. Academic Press, New York (1973)
17. BOULTER, D., JEREMY, J. J. and WILDING, M. Amino acids liberated into the culture medium by pea seedling roots, *Plants and Soil*, **24**, 121–127 (1966)
18. BRAMS, E. The mucilaginous layer of citrus roots—its delineation in the rhizosphere and removal from roots, *Plant and Soil*, **30**, 105–108 (1969)

19. BROWN, M. E. Seed and root bacterization, *Annual Review of Phytopathology,* **12,** 311–331 (1974)

20. CAMPBELL, R. and ROVIRA, A. D. The study of the rhizosphere by scanning electron microscopy, *Soil Biology and Biochemistry,* **5,** 747–752 (1973)

21. CLOWES, F. A. L. Apical meristems of roots, *Biological Reviews,* **34,** 501–529 (1959)

22. CLOWES, F. A. L. and JUNIPER, B. E. *Plant Cells.* Blackwell, Oxford (1968)

23. CORMACK, R. G. H. The development of root hairs in angiosperms, *Botanical Review,* **15,** 583–612 (1949)

24. CORMACK, R. G. H. Development of root hairs in angiosperms. II, *Botanical Review,* **28,** 446–464 (1962)

25. COX, G. and SANDERS, F. Ultrastructure of the host–fungus interface in a vesicular–arbuscular mycorrhiza, *New Phytologist,* **73,** 901–912 (1974)

26. CRUSH, J. R. Plant growth responses to vesicular–arbuscular mycorrhiza. VII. Growth and nodulation of some herbage legumes, *New Phytologist,* **73,** 743–749 (1974)

27. CURRIER, H. B. Callose substance in plant cells, *American Journal of Botany,* **44,** 478–488 (1957)

28. DARBYSHIRE, J. F. and GREAVES, M. P. The invasion of pea roots, *Pisum sativum* L., by soil microorganisms *Acanthamoeba palestinensis* (Reich) and *Pseudomonas* sp., *Soil Biology and Biochemistry,* **3,** 151–155 (1971)

29. DARBYSHIRE, J. F. and GREAVES, M. P. Bacteria and protozoa in the rhizosphere, *Pesticide Science,* **4,** 349–360 (1973)

30. DART, P. J. Scanning electron microscopy of plant roots, *Journal of Experimental Botany,* **22,** 163–168 (1971)

31. DART, P. J. and MERCER, F. V. The legume rhizosphere, *Archiv für Mikrobiologie,* **47,** 344–378 (1964)

32. DE ROTHSCHILD, D. I. Nodulación en leguminosas subtropicales de la flora argentina, *Revista del Museo Argentino de Ciencias Naturales 'Bernardino Rivadavia',* **3,** 367–386 (1970)

33. DÖBEREINER, J., DAY, J. M. and DART, P. J. Nitrogenase activity and oxygen sensitivity of the *Paspalum notatum-Azotobacter paspali* associations, *Journal of General Microbiology,* **71,** 103–116 (1972)

34. ESAU, K. *Plant Anatomy.* Wiley, New York (1953)

35. FLOYD, R. A. and OHLROGGE, A. J. Gel formation on nodal root surfaces of *Zea mays.* I. Investigation of the gel's composition, *Plant and Soil,* **33,** 331–343 (1970)

36. FLOYD, R. A. and OHLROGGE, A. J. Gel formation on nodal root surfaces of *Zea mays.* Some observations relevant to understanding its action at the root–soil interface, *Plant and Soil,* **34,** 595–606 (1971)

37. FOSTER, R. C. and ROVIRA, A. D. The rhizosphere of wheat roots studied by electron microscopy of ultrathin sections, *Bulletins from the Ecological Research Committee* (Stockholm), **17,** 93–102 (1973)

38. FRANKE, W. Mechanism of foliar penetration of solutions, *Annual Review of Plant Physiology,* **18,** 281–300 (1967)

39. FREY-WYSSLING, A. and MÜHLETHALER, K. *Ultrastructural Plant Cytology,* 326. Elsevier, Amsterdam (1965)

40. GARRETT, S. D. *Pathogenic Root-infecting Fungi.* Cambridge University Press (1970)

41. GREAVES, M. P. and DARBYSHIRE, J. F. The ultrastructure of the mucilaginous layer on plant roots, *Soil Biology and Biochemistry,* **4,** 443–449 (1972)

42. HARLEY, J. L. *Biology of Mycorrhiza,* 2nd edition. Leonard Hill, London (1969)

43. JENNY, H. and GROSSENBACHER, K. Root–soil boundary zones as seen in the electron microscope, *Soil Science Society of America Proceedings,* **27,** 273–277 (1963)

44. JUNIPER, B. E. and ROBERTS, R. M. Polysaccharide synthesis and the fine structure of root cap cells, *Journal of the Royal Microscopical Society*, **85**, 63–72 (1966)
45. KASPARI, H. Elektronenmikroskopische Untersuchung zur Feinstruktur der endotrophen Tabakmykorrhiza, *Archiv für Mikrobiologie*, **92**, 201–207 (1973)
46. KHAN, A. G. Podocarpus root nodules in sterile culture, *Nature*, **215**, 1170 (1967)
47. KHAN, A. G. Podocarp-type mycorrhizal nodules in *Aesculus indica*, *Annals of Botany*, **36**, 229–238 (1972)
48. KHAN, A. G. and VALDER, P. G. The occurrence of root nodules in the Ginkgoales, Taxales and Coniferales, *Proceedings of the Linnean Society of New South Wales*, **97**, 35–41 (1972)
49. KESSLER, K. J. Growth and development of mycorrhizae on sugar maple (*Acer saccharum* Marsh), *Canadian Journal of Botany*, **44**, 1413–1425 (1966)
50. LAMBERTZ, P. Untersuchungen über das Vorkommen von Plasmodesmen in den Epidermisaussenwänden, *Planta*, **44**, 147–190 (1954)
51. LANGE, R. T. Additions to the known nodulating species of Leguminosae, *Antonie van Leeuwenhoek Journal of Microbiology and Serology*, **25**, 272–276 (1959)
52. LEISER, A. T. A mucilaginous root sheath in Ericaceae, *American Journal of Botany*, **55**, 391–398 (1968)
53. MACGREGOR, A. N. and ALEXANDER, M. Formation of tumor-like structures on legume roots by *Rhizobium*, *Journal of Bacteriology*, **105**, 728–732 (1971)
54. MARCHANT, R. The root surface of *Ammophila arenaria* as a substrate for microorganisms, *Transactions of the British Mycological Society*, **54**, 479–506 (1970)
55. MORRÉ, D. J., JONES, D. D. and MOLLENHAUER, H. H. Golgi apparatus mediated polysaccharide secretion by outer root cap cells of *Zea mays*, *Planta*, **74**, 286–301 (1967)
56. MOSSE, B. Advances in the study of vesicular–arbuscular mycorrhiza, *Annual Review of Phytopathology*, **11**, 171–196 (1973)
57. MOSSE, B. Plant growth responses to vesicular–arbuscular mycorrhiza. IV. In soil given additional phosphate, *New Phytologist*, **72**, 127–136 (1973)
58. MOSSE, B. and HEPPER, C. Vesicular–arbuscular mycorrhizal infections in root organ cultures, *Physiological Plant Pathology* (in press) (1975)
59. NEAL, J. L., TRAPPE, J. M., LU, K. C. and BOLLEN, W. B. Some ectotrophic mycorrhizae of *Alnus rubra*, in *Biology of Alder:* Proceedings of the Northwest Scientific Association. 40th Annual meeting (edited by J. M. Trappe, J. F. Franklin, R. F. Tarrant and G. M. Hansen), 179–184 (1968)
60. RODRIGUEZ-BARRUECO, C. and BERMUDEZ DE CASTRO, F. Cytokinin-induced pseudonodules on *Alnus glutinosa*, *Physiologia Plantarum*, **29**, 277–280 (1973)
61. ROELOFSEN, P. A. The plant cell-wall, in *Encyclopedia of Plant Anatomy* (edited by W. Zimmermann and P. G. Ozenda), 2nd edition. Gebrüder Borntraeger, Berlin (1959)
62. ROSENE, H. F. A comparative study of the rates of water influx into the hairless epidermal surface and the root hairs of onion roots, *Physiologia Plantarum*, **7**, 676–686 (1954)
63. ROVIRA, A. D. Plant root exudates, *Botanical Review*, **35**, 35–57 (1969)
64. ROVIRA, A. D. Zones of exudation along plant roots and spatial distribution of microorganisms in the rhizosphere, *Pesticide Science*, **4**, 361–366 (1973)
65. SCANNERINI, S. and BELLANDO, M. 'Sull' ulstrastuttura delle microrrize endotrofiche di *Ornithogallum umbellatum* L. in attivita vegetativa, *Atti delle Accademia delle Scienze Torino*, **102**, 795–809 (1968)
66. SCOTT, F. M. The anatomy of plant roots, in *Ecology of Soil-borne Plant Pathogens* (edited by K. F. Baker and W. C. Snyder), 145–151. University of California Press, Berkeley (1965)

67. SCOTT, F. M., BYSTROM, B. G. and BOWLER, E. Root hairs, cuticle and pits, *Science,* **140,** 63–64 (1963)
68. SLANKIS, V. The role of auxin and other exudates in mycorrhizal symbiosis of forest trees, in *Physiology of Forest Trees* (edited by K. V. Thimann). Ronald Press, New York (1958)
69. SLANKIS, V. Hormonal relationships in mycorrhizal development, in *Ectomycorrhizae* (edited by G. C. Marks and T. T. Kozlowski), 232–298. Academic Press, New York (1973)
70. SUN, C. N. Growth and development of primary tissues in aerated and non-aerated roots of soybean, *Bulletin of the Torrey Botanical Club,* **82,** 491–502 (1955)
71. TALBOYS, P. W. Some mechanisms contributing to *Verticillium*-resistance in the hop root, *Transactions of the British Mycological Society,* **41,** 227–241 (1958)
72. TRAPPE, J. M. Tuberculate mycorrhizae of Douglas fir, *Forest Science,* **11,** 27–32 (1965)
73. TRINICK, M. J. Symbiosis between *Rhizobium* and the non-legume *Trema aspera, Nature,* **224,** 459–460 (1973)
74. VOZZO, J. A. and HACSKAYLO, E. Anatomy of mycorrhizae of selected eastern forest trees, *Bulletin of the Torrey Botanical Club,* **91,** 378–387 (1964)
75. WOOD, R. K. S. *Physiological Plant Pathology.* Blackwell, Oxford (1967)
76. ZAK, B. Classification of ectomycorrhizae, in *Ectomycorrhizae* (edited by G. C. Marks and T. T. Kozlowski), 43–78. Academic Press, New York (1973)

4 Phosphorus Cycling by Soil Micro-organisms and Plant Roots

D. S. HAYMAN

Phosphorus is a major element in plant nutrition and plants obtain it from soil. Factors that affect the solubility of soil phosphates, their uptake by plants and their return to the soil in plant residues are of fundamental importance to plant growth. The role of micro-organisms in these processes is discussed in this chapter.

4.1 The phosphorus cycle and ecosystems

The turnover of soil phosphorus is traditionally represented as a cycle (*Figure 4.1*). One major feature of this cycle is that the bulk of the phosphorus in soil is 'hors de combat' and only about 1% of it or less is incorporated into the above-ground vegetation during a growing season. This applies both to high-yielding agricultural crops and to natural stands of vegetation (*Table 4.1*). With agricultural crops the phosphorus cycle is an open one because of the input of fertiliser phosphate to replace that permanently removed in the harvest. With natural vegetation the cycle is virtually closed, except for very small amounts that enter in rainfall and from the deeper soil or are lost by leaching, and most plant phosphate is recycled by microbial breakdown of litter and organic debris. A vivid example of this is the Brazilian rain forest, where lush vegetation overlies an impoverished soil which cannot support good crop growth after removal of the forest.[81] Of the 6.9 and 9.4 kg P/ha taken up annually

Table 4.1
ANNUAL TURNOVER OF PHOSPHORUS IN DIFFERENT ECOSYSTEMS

Vegetation	Age (yr)	Country	Total P in soil (kg/ha)	Annual uptake of P by vegetation (kg/ha)	Reference
Barley	1	England	2 200* (top 25 cm)	10	Cooke[22]
Potatoes	1	England	2 050† (top 25 cm)	11	Cooke[22]
Average crop in 1930	1	USA	1 550 (top 25 cm)	6	Black[15]
Native grassland		Canada	2 913 (top 30 cm)	2.1	Halm, Stewart and Halstead[39]
Heather heath	2	Scotland	437	2	Robertson and Davies[69]
Oak forest	70	Belgium	920 (top 40 cm)	6.9	Duvigneaud and Denaeyer-Desmet[27]
Oak–ash forest	140	Belgium	2 200 (top 40 cm)	9.4	Duvigneaud and Denaeyer-Desmet[27]
Douglas fir forest	36	USA	2 362 (top 30 cm)	7.2	Cole, Gessel and Dice[21]

* Estimated from 735 ppm total P.
† Estimated from 682 ppm total P.

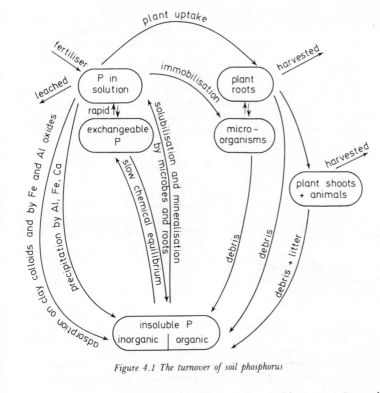

Figure 4.1 The turnover of soil phosphorus

by the Belgian forests referred to in *Table 4.1*, 4.7 and 5.4 kg P/ha^{-1} year^{-1}, respectively, were returned to the forest floor, mainly in litter. Also, much of the phosphorus retained by the trees is recycled within them: for example, 3.3 out of the 5.5 kg P/ha taken up by a stand of *Pinus taeda* in its twentieth year.[85]

The distribution of phosphorus in the Canadian grassland cited in *Table 4.1* was examined in detail. In July, expressed as kg P/ha, the green plant material contained 2.1, the standing dead plant material 2.1, the litter 0.7, the roots and rhizomes 10.7, consumers <1.0, and the soil fauna and micro-organisms 19.8. Thus, the soil micro-organisms apparently contain at least as much phosphorus per hectare as all the vegetation.

4.2 Phosphorus compounds in soil

Since only about 1% of the phosphorus in any ecosystem is found in the living components, some discussion of the other 99% is required as a background to the activities of plants and microbes. The chemistry of soil phosphate is complex and full details can be consulted elsewhere.[15, 19, 22, 50, 72, 76, 83] Here we are concerned with generally accepted principles and facts.

Soil phosphate is most conveniently divided into two categories—insoluble and readily soluble. The insoluble phosphate, which is not directly available to plants or micro-organisms, usually comprises around 95–99% of the total soil phosphate (*Table 4.2*).

Table 4.2

PHOSPHORUS CONTENTS OF
A RANGE OF SOILS

Soil	*Total* P (ppm)	0.5 M $NaHCO_3$- *soluble* P (ppm)	0.01 M $CaCl_2$- *soluble* P (μmol/l)
Rothamsted (Barnfield, No P)	610	17	0.5
Rothamsted (Barnfield, +PKNaMg)	1170	99	7.3
Ashridge (arable fallow)	720	16	1.2
Woburn (arable fallow)	1552	26	1.4
Bower Heath (grassy common)	780	18	1.0
Wareham (heath)	132	5	3.2
St. Leonard's (heath)	288	8	0.7
Millbrook (deciduous forest)	1048	10	0.8

The first two soils are cited by Cooke[22], the last six by Hayman and Mosse.[44]

The insoluble inorganic phosphate in all soils is mostly attached to three elements, of which iron and aluminium are the main binding agents in acid soils and calcium in slightly acid to alkaline soils. Their amounts can be estimated by using a series of chemical extractants.[20] Most of the inorganic phosphate in the earth's crust is apatite (rock phosphate), principally insoluble calcium fluorapatite, $3Ca_3(PO_4)_2 \cdot CaF_2$. This weathers in soil to yield several secondary minerals such as hydroxy-apatite, $3Ca_3(PO_4)_2 \cdot Ca(OH)_2$. The large mass of insoluble iron and aluminium compounds in soil includes hydrated phosphates such as strengite, $FePO_4 \cdot 2H_2O$, and variscite, $AlPO_4 \cdot 2H_2O$. In acid soils, phosphate is precipitated on the surfaces of iron and aluminium oxides or by aluminium ions free in solution or attached to silicate crystals such as kaolinite and montmorillonite. These processes are sometimes called 'phosphate fixation'. Adding lime to acid soils causes the OH^- ion to exchange with the $H_2PO_4^-$

ion in the hydroxy phosphate, thereby increasing the phosphate in solution.

The amount of the total insoluble soil phosphorus which is organic varies widely—around 30–85%, discounting extreme types of soils.[4, 24] It is particularly high in acid soils. Usually it is related to the amounts of carbon and nitrogen in the soil in the proportions of about 110 parts carbon to 9 parts nitrogen to 1 part phosphorus. About 30–50% of it consists of phytin material, largely inositol phosphates. Up to 3% is in nucleic acids and nucleotides, and about 1% in phospholipids. There are also traces of sugar phosphates and lower inositol phosphates. This leaves at least half of the total soil organic phosphorus unaccounted for, but it may occur in complexes with humic material. Phytin accumulates mainly as insoluble iron and aluminium phytates in acid soils and as insoluble calcium phytates under alkaline conditions. Nucleic acids are more readily broken down and this explains in part their small quantity in soils; their decomposition is somewhat slowed down through adsorption on clay particles.

The insoluble soil phosphate only reaches equilibrium very slowly with the readily soluble fraction. There is no sharp boundary between these fractions because most of the soluble phosphate ions are adsorbed by the solid phase. These adsorbed ions rapidly exchange with the phosphate ions in the soil solution. The amount of this readily soluble phosphate present on adsorption sites and in the soil solution is often referred to as exchangeable phosphate, plant-available phosphate or the labile pool. This pool is replenished by desorption of phosphate by other ions (e.g. HCO_3^-, silicate, OH^-) and by the solution of insoluble inorganic phosphates and mineralisation of organic phosphate. It is best measured by equilibrating soil with a solution containing radioactive ^{32}P and growing plants in this labelled soil. The ratio of ^{32}P to ^{31}P (specific activity) in the plants (used to calculate the 'L'-value) reflects the ratio of ^{32}P to ^{31}P in the labile pool of soil phosphate (used to calculate the 'E'-value). Thus, if there is much exchangeable phosphate in the soil, the specific activity of the phosphorus in the pool will be low. Some reagents, e.g. $NaHCO_3$, rapidly desorb phosphate and provide an approximate measurement of part of the 'pool' without dissolving less available phosphates (*Table 4.2*).

The relationship between amounts of phosphate on the exchangeable sites and in the soil solution expresses the capacity of the soil to supply soluble phosphate. In a soil with a small adsorption capacity the concentration in the soil solution will decrease rapidly, even if it were high initially, as plants grow in that soil, whereas it will change little during crop growth if it is sustained by a large exchangeable pool.

Only a small amount of readily soluble phosphate is present in the soil solution at any one time. Its amount is commonly estimated in mild extractants such as $CaCl_2$ solution (*Table 4.2*). Concentrations of 10^{-7} M or less occur in poor soils, but 10^{-6} M (equivalent to 0.03 ppm P) is more common. Concentrations may reach 10^{-5} or 10^{-4} M in some soils given fertiliser. In dilute solution the form of phosphate ion varies according to pH and phosphoric acid dissociates thus:

$$H_3PO_4 \rightleftharpoons H^+ + H_2PO_4^- \rightleftharpoons H^+ + HPO_4^{2-} \rightleftharpoons H^+ + PO_4^{3-}$$

At pH 6.0, 94% of the phosphate is present as the $H_2PO_4^-$ ion; and at pH 7.0, 61%. Between pH 5 and pH 9 the amounts of undissociated H_3PO_4 and of the trivalent PO_4^{3-} ion are negligible.

4.3 Plant uptake of soil phosphate

It is generally agreed that most, if not all, of the soil phosphorus taken up by plants is in the form of orthophosphate ions in the soil solution. Although some organic phosphate is present in the soil solution,[90] whether this is taken up directly by plants needs further study. Plants are generally said to 'prefer' the $H_2PO_4^-$ ion to the HPO_4^{2-} ion. Since the former is the dominant ion at the pH at which 'fixation' of phosphate is least, i.e. pH 6.5, most plant-available phosphate is normally present at this pH.

The average orthophosphate concentration in the soil solution of around 10^{-6} M is near the limit at which plants can absorb adequate phosphate. However, critical (threshold) concentrations vary for different plants. For example, Rorison,[70] using solution cultures, found that *Urtica dioica,* which only grows in reasonably fertile soils, thrived only at 10^{-5} M or above, whereas *Rumex acetosa,* which has a broad edaphic tolerance, survived over a whole range of concentrations. *Scabiosa columbaris* grew poorly at 10^{-7} M and best at 10^{-3} M. In contrast, *Deschampsia flexuosa* grew continuously at concentrations down to 10^{-7} M and responded only slightly to 10^{-3} M. These threshold concentrations seem to be related to the inherently different growth rates and metabolism of different plant species. Much metabolic energy is in fact needed to absorb phosphate, because the concentration inside the plant root is usually over 1000 times greater than in the soil solution, values of 10^{-2} M in the cortical cells and 10^{-3} M in the xylem sap being suggested.[14]

An average crop removing 10 or more kg P/ha^{-1} year^{-1} requires hundreds of times as much phosphorus as can be found at any one time in the soil solution within the rooting depth, the latter being of the order of 0.04 kg P/ha in a typical agricultural soil (see reference 72). Thus, the most vital factor affecting plant uptake of soil phosphate is the rate at which an adequate concentration of phosphate is maintained in the soil solution at the root surface. In normal soils this is not achieved, because, although desorption of labile phosphate into the soil solution is very rapid for the soil mass as a whole, roots are only in contact with a relatively small volume of soil. The labile phosphate therefore becomes depleted at the root surface, although not completely,[6, 29] and must be replenished by soluble phosphate from the soil further away. However, phosphate ions are not very mobile in soil. They move chiefly by diffusion through the soil solution down a concentration gradient induced by the lowered concentration of ions at the root surface. Diffusion of phosphate ions is retarded by the fine soil particles which make their diffusion path more tortuous and reversibly bind them to their surfaces. Movement of phosphate ions by mass flow in the water absorbed and transpired by plants can only account for a small proportion of the phosphate absorbed. Phosphate is thus replenished in the soil solution close to the root far more slowly than plants take it up. This results in a zone of phosphate depletion around the root. This zone is about 1 or 2 mm wide and can be seen in autoradiographs of labelled soil around roots.[13, 51] It coincides with the rhizosphere (see Chapter 2)—the region around roots where micro-organisms are particularly active and therefore influence root–soil interactions.

The ability of some plants, such as lupins, to obtain much more phosphorus than other species growing in the same soil might at first sight suggest that they can tap more of the insoluble phosphate. This does not appear to be so, however, because experiments in ^{32}P-labelled soil[66, 67] have shown that different plant species take up phosphate from the same source and differences in total P uptake are due to some species tapping more from the labile pool rather than dissolving more of the insoluble phosphate. The ^{32}P/^{31}P ratio was the same for the different species, whereas it would have been lower in any species capable of utilising some of the differently labelled insoluble phosphate.

Addition of phosphate fertiliser to soil disturbs the equilibria discussed so far and maintains high concentrations of soluble phosphate near the root for long periods. However, only about 25% of the phosphate fertiliser is taken up by the crop in its year of application. The rest builds up as 'residual phosphate', reaching 40% of the total phosphate in cultivated soils in Britain,[22] and remains more available to crops grown in subsequent years, and probably to micro-organisms, than is the native insoluble soil phosphate. In natural ecosystems soil phosphate equilibria are disturbed in the springtime during the flush of litter breakdown.

4.4 Mechanisms of microbial transformations of phosphorus compounds

The literature on this subject is vast; therefore only a few references will be cited in this outline of the four generally recognised processes involved.

Solubilisation of inorganic phosphate

Organic acids. The chief way that micro-organisms dissolve insoluble phosphates, suspended as fine particles in agar media, is by secreting organic acids such as lactic, glycolic, oxalic and citric. Clear zones develop around the microbial colonies. This method, used nearly 70 years ago to detect and isolate phosphate-dissolving bacteria from soil,[73] is fully described by Louw and Webley,[52] and depends on metabolism of a substrate such as glucose. Bacteria and fungi in such cultures can dissolve enough phosphate from, for example, $Ca_3(PO_4)_2$ for their own growth and release surplus orthophosphate ions. Hydroxy acids are better able to dissolve apatite than are other acids,[84] presumably because they chelate with calcium in addition to lowering the pH. Some bacteria produce 2-keto-gluconic acid,[26] which also chelates with calcium. This dissolves more calcium, so restoring the chemical equilibrium, and this is balanced by phosphate ions coming into solution at the same time, which results in an increase of orthophosphate in solution. Further evidence that chelation may be more important than simple pH effects is that oxalic acid dissolves more phosphate at pH 4.4 than gluconic acid at pH 4.3.[56]

Inorganic acids. In the special case of chemo-autotrophs that can oxidise ammonium or sulphur, nitric or sulphuric acids are produced. These can release orthophosphate ions from rock phosphate.

Carbon dioxide. Micro-organisms can lower the pH around them by the CO_2 given off during respiration thereby causing more phosphate to dissolve.

Hydrogen sulphide. Certain bacteria release H_2S, which can react with ferric phosphate to yield ferrous sulphide and soluble orthophosphate ions.[79]

Humic substances. 'Humic acid' and 'fulvic acid' from microbially degraded plant debris may chelate with the calcium, iron or aluminium in complex phosphates, so releasing orthophosphate. They can also form stable, soluble complexes with iron or aluminium and phosphate which are accessible to plant roots.[56]

Mineralisation of organic phosphate

Phytases. Phytases hydrolyse the phosphate ester linkage of soluble salts of phytic acid to yield inositol and orthophosphate. When Greaves and Webley[37] added sodium *myo*-inositol hexaphosphate to sand containing soil micro-organisms, 72% of the substrate phosphate was released at pH 6.8. However, the less soluble calcium salt and the sparingly soluble magnesium, iron and aluminium salts in the same conditions released only 2.6, 3.0, 0.1 and 0.3%, respectively. *Aerobacter aerogenes* hydrolysed *myo*-inositol polyphosphate in a step-wise manner to yield several intermediate polyphosphates, but was inactive against the *scyllo* isomer.[34]

Nucleases. Nucleic acids are rapidly degraded. For example, *Cytophaga johnsonii* released a mixture of nucleotides after 24 h and much orthophosphate during 10 days from RNA and DNA in liquid culture.[35]

Phospholipases. The commonest phospholipids in soil, such as lecithin (phosphatidyl choline) and phosphatidyl ethanolamine, release orthophosphate through enzyme hydrolysis.

Organic acids. Organic phosphates can be dissolved by bacterial organic acids, after which free phosphate may sometimes be liberated by hydrolysis.[10]

Immobilisation

The incorporation of soluble phosphate into growing microbial cells

75

is referred to as immobilisation. In solution cultures micro-organisms can decrease phosphate transfer to plant tops at low (<1 ppm) but not high (10 ppm) concentrations of phosphorus by incorporating soluble phosphate into their own cells, especially in nucleic acids.[8,9] This temporarily immobilised phosphate is released when the cells die and may then be available to plants again. Indeed plants grown in sand containing rock phosphate took up much more phosphate from dead bacteria that had grown in this medium than from the original rock phosphate.[87]

Bacteria and actinomycetes accumulate rather more phosphorus (1.5–2.5% of their dry weight) than fungi (0.5–1%) or plants (0.05–0.5%).

Oxidation and reduction of inorganic phosphate

Orthophosphate is the most oxidised state of phosphorus. Some heterotrophic micro-organisms can assimilate phosphite, HPO_3^{2-}, and convert it to phosphate. Conversely, *Clostridium butyricum* may convert phosphate to phosphite and hypophosphite in very wet soils.[89] Adams and Conrad[1] suggested a microbial conversion of phosphite to phosphate in soil.

These oxidation–reduction reactions are not very important in soil and will not be discussed further.

4.5 Microbial solubilisation, mineralisation and immobilisation of phosphate in soil

The halcyon days of a microbe in a Petri dish bear no relationship to the turbulent times it endures in soil. In pure culture there is normally abundant substrate, optimum temperature, no antagonism from other microbes and no adsorption by clay particles, whereas in soil most of the microbial population is forced into dormancy. The appearance of a piece of substrate in soil unleashes a flood of activity in the microhabitat around it, lasting only until it is used up. The mechanisms described in Section 4.4 must be considered in this context, since substrate availability will broadly determine where and how micro-organisms in the soil increase or deplete the pool of labile phosphate to an extent that can benefit or inhibit plant uptake of phosphate.

Solubilisation

Many common micro-organisms, including species of *Pseudomonas, Achromobacter, Flavobacterium, Streptomyces,* and especially *Aspergillus* and *Arthrobacter,* can dissolve insoluble inorganic phosphates known to occur in soil.[2, 80, 84, 86] A high proportion of the total microbial population has this ability, up to 85% in some soils,[49] although many lose it on subculturing. The rhizosphere sometimes has a particularly high proportion of such organisms, Swaby and Sperber,[84] for example, having found that 20–40% of the bacteria, actinomycetes and fungi isolated from the rhizospheres of many plants were able to dissolve hydroxyapatite, compared with 10–15% from non-rhizosphere soil. Sometimes there is no preferential stimulation of phosphate-solubilising bacteria by roots.[46]

Solubilisation must be considered within the rhizosphere, because beyond this region, which coincides with the zone of phosphate depletion, solubilisation would have little effect on the growing root because of the slow diffusion of phosphate ions in soil (see Section 4.3). Root exudates and sloughed cells provide the substrate to support the intense microbial activity characteristic of the rhizosphere. Therefore, colonies of bacteria could dissolve phosphates at sites of exudation along the root by producing organic acids and CO_2—compounds also excreted by the roots themselves. Organic acids in root exudates can also be adsorbed by clay minerals, thereby decreasing the number of sites able to adsorb phosphate and so keeping more phosphate ions in solution.[64] This mechanism may well apply to soil micro-organisms too. However, the quantities of organic acids from roots may be so much greater than those from micro-organisms that they will mask the microbial effects on pH and adsorption.

Another mechanism which is applicable to roots and which may be true of micro-organisms is the uptake of calcium ions causing a shift in the mass action equilibria of sparingly soluble calcium phosphates to bring more calcium and therefore phosphate ions into solution. Johnston and Olsen[45] grew plants with different feeding powers in solution culture and calculated that calcium adsorption and/or absorption by the roots accounted for 75% or more of the phosphate brought into solution, whereas excretion by roots of CO_2, chelates or acids had only negligible effects. Such a dissolution process would be favoured by close contact with the surfaces of mineral particles, and this is more readily achieved by bacterial cells and fungal hyphae, because of their small size, than by plant roots.

77

Indeed fungi obtain more phosphate when their hyphae are in contact with particles of finely ground minerals than when contact is interrupted by a paper disc between them.[84]

Even though orthophosphate may be brought into the labile pool by organic acids or calcium uptake in the rhizosphere, there is such vigorous competition for substrate (root exudate) by the rhizosphere micro-organisms that their growth and demand for phosphate causes them to compete effectively with roots for any phosphate liberated, and so the roots may not gain much.

The ability of *Thiobacillus* spp. to oxidise sulphur to sulphuric acid is used commercially in the Lipman process. Here a mixture of soil, manure, elemental sulphur and rock phosphate is prepared and the acid produced by the bacteria dissolves the phosphate. However, this process is not as cheap or as efficient as the use of standard phosphate fertilisers, except perhaps in warm, wet soils.

Under anaerobic conditions in waterlogged soils many bacteria produce hydrogen sulphide, which may increase the availability of iron phosphates by converting them to black ferrous sulphide and liberating phosphoric acid. This may explain the greater availability of phosphate in waterlogged rice fields.

Good evidence that free-living micro-organisms can solubilise phosphate in soils in sufficient amounts to benefit the plant comes from experiments in which the addition of straw to soil increased phosphate uptake both from added insoluble salts and from sparingly soluble native soil phosphate.[56] The microbially degraded straw released 'humic' and 'fulvic' acids which formed stable complexes with iron and phosphate which were as available to the plant as orthophosphate ions.

Although microbial solubilisation of inorganic phosphates may not occur on a large scale in most soils, there are certain soil types where it is more likely:[84] (1) acid, sandy soils for acid-tolerant plants, e.g. lupins; (2) old pastures where micro-organisms and humus are plentiful; and (3) light wet soils, low in sesquioxides but high in sulphates and organic matter. Also, it is feasible that soil microbes could release more phosphate from added rock phosphate or the adsorbed residual fertiliser phosphate than from the more firmly bound native phosphate.

Brief mention should be made here of inoculation of plants with phosphate-solubilising bacteria to obtain yield increases in the field. These so-called 'bacterial fertilisers' or 'phosphobacterin' are reported to increase crop yields in the Soviet Union[23, 57] and Spain[5] but not in the United States.[78] The last-mentioned lack of response is not really surprising, because phosphate-solubilising bacteria are common in most soils. Brown,[18] in her review on root and seed bacterisation, suggested that any stimulation of plant growth by inoculating seeds with bacteria is more likely to be due to growth hormones produced by the bacteria than to appreciable phosphate solubilisation. Increased uptake of phosphate by plants inoculated with soil micro-organisms was clearly demonstrated in sand given poorly soluble inorganic phosphates and phosphate-free nutrient solution.[31] However, the same process does not necessarily happen in soil. Also, the appearance of the plants, particularly their long internodes, is suggestive of growth stimulation by bacterial hormones,[18] which itself could lead to more phosphate uptake.

Mineralisation

Some species, such as *Bacillus megaterium,* can release orthophosphate from both organic and inorganic phosphates. Many soil micro-organisms can mineralise complex organic phosphates, including species of *Arthrobacter, Proteus, Serratia, Streptomyces, Aspergillus* and *Rhizopus.* A high percentage of the soil micropopulation has this ability, up to 70–80% in extreme cases,[49] but just under 50% is a more typical example.[33] In a systematic survey of organic phosphate-decomposing organisms on grass root surfaces and in rhizosphere and non-rhizosphere soil, Greaves and Webley[36] found that about 70% produced phosphatases that hydrolysed phenolphthalein phosphate and about 40, 20, 40, 40 and 30% produced glycerophosphatase, lecithinase, phytase, ribonuclease and deoxyribonuclease, respectively. Preferential stimulation of phosphatase producers by roots was infrequent, and occurred mainly at anthesis.

Although many organisms show phytase activity *in vitro,* most phytate in soil is not in solution but in sesquioxide gels of iron and aluminium or adsorbed by clay minerals where it is not accessible to enzymes. Consequently, phytate accumulates in soil and forms a much higher percentage of the soil organic phosphate than of the organic phosphate in debris entering soil. For example, although sodium phytate is readily hydrolysed in sand by soil micro-organisms (see p. 75), adding soil or clay minerals inhibits hydrolysis.[37] With aluminium montmorillonite both adsorption and inactivation of the enzyme seemed to occur, whereas sodium montmorillonite only adsorbed it.

Microbial degradation of nucleic acids in soil must occur much faster than the breakdown of phytates, because less than 3% of the soil organic phosphorus is nucleic acid compared with up to 10% in the organic residues entering soil. The DNA which persists seems to be of microbial origin and its amount is probably greater than can be accounted for in bacterial cells present in soil.[4] Some nucleic acid must be protected from the action of microbial nucleases, possibly by adsorption by clay minerals such as montmorillonite .[38]

The production of phosphatase enzymes by roots themselves is important in the nutrition of plants in soil,[91] and is probably more important than phosphatase production by rhizosphere organisms. For example, Estermann and McLaren[28] found that non-sterile barley roots liberated only little more orthophosphate from β-glycerol phosphate than did sterile-grown barley roots. Similarly, Ridge and Rovira[68] assayed phosphatase activities of wheat roots under sterile and non-sterile conditions, and concluded that most enzyme activity was associated with the root surface and was not much increased by rhizoplane micro-organisms. Inoculation of roots with pure cultures of bacteria and a fungus known to have phosphatase activity did not increase the activity of the roots. When Martin[53] supplied wheat plants grown in nutrient solution or in soil with ^{32}P-labelled inositol hexaphosphate at concentrations considered typical of the soil solution (0.2 ppm P), he found that incorporation of ^{32}P into the plants due to enzyme activity was not increased by inoculation with phytase-producing bacteria or a mixed rhizosphere microflora.

The over-all impression from the extensive literature on the subject is that microbial mineralisation of organic phosphate in soil, particularly certain compounds, occurs widely. However, except where microbes find Utopia as in litter in springtime, the liberation of soluble phosphate surplus to their own requirements probably contributes only slightly to the phosphorus nutrition of plants.

Immobilisation

Soil micro-organisms compete with roots for phosphate from the labile pool. In normal cultivated land the bacteria alone are estimated to take up and thus immobilise 4 to 10 kg P/ha.[76] Adding to this fungal and actinomycete uptake, it would seem that micro-organisms remove at least as much phosphorus per hectare as the crop (see Section 4.1). The phosphate immobilised within microbial cells is released when they die. Whether other microbes capture most of this released phosphate in soil, thereby operating a 'mini-cycle' that almost completely by-passes the roots, or whether the roots benefit as in sand (see reference 87) needs to be investigated.

During decomposition of plant residues in soil there is an increase in the microbial population and, hence, a greater demand for phosphate. If the carbon–phosphorus ratio of the residues is high, micro-organisms may assimilate labile phosphate to the extent that yields of crops growing in that soil will be depressed unless compensated for by applications of fertiliser. There is a critical concentration of about 0.2% phosphorus[3] in the organic matter being decomposed, above and below which there will be net mineralisation or immobilisation, respectively.

Results from pot experiments[61] suggest that microbial activity may deplete the pool of labile phosphate in a woodland soil. The phosphorus in mycorrhizal plants grown in this soil labelled with ^{32}P had the same specific activity irrespective of whether the soil had been sterilised. If the indigenous microbial population had mobilised significant amounts of the unlabelled organic phosphate, relatively more ^{31}P than ^{32}P would have been taken up from the unsterile soil. Furthermore, the plants in the sterile soil had taken up two or three times as much phosphorus. This suggests not only effects of pathogens or root inhibitors in the unsterile soil but also a shortage of available phosphate caused by competition by plant roots and the indigenous microbial population for the same pool of labile phosphorus.

4.6 Effects of micro-organisms on plant roots

Micro-organisms affect root morphology. Bowen and Rovira,[17] for example, observed in sand cultures a 25–40% inhibition by micro-organisms of root growth of subterranean clover, tomato, *Phalaris* and *Pinus radiata,* and a 20–53% reduction in total number of lateral roots, also the inhibition on agar of root hair growth and development in subterranean clover. These effects would be of great importance in soil where the volume of soil explored is affected both by growth of roots themselves and growth of root hairs which extend the zone of phosphate depletion.

Plant metabolism and, hence, phosphate uptake can be directly affected by micro-organisms. This explained in part the translocation of 4.4 times as much ^{32}P to the tops of tomato seedlings grown in solution culture in non-sterile compared with sterile conditions.[16] Since the ^{32}P reached the shoots within 20 min, the increased uptake could not be attributed solely to absorption by the rhizoplane microflora. The possible effect of bacterial hormones on plant uptake of phosphate are referred to above (p. 79).

4.7 Mycorrhiza

Mycorrhizal fungi (see reference 40), being symbiotic organisms, have an ecological niche with abundant substrate inside plant roots from which their hyphae extend into the soil. Plants without mycorrhiza, which are not often found in natural or cultivated soils, generally take up much less phosphate from soils deficient in labile phosphate than plants with mycorrhiza.

Ectotrophic mycorrhiza

The ectotrophic (or ecto-) mycorrhizal fungi are associated with the roots of many tree species. Morphological changes in roots induced by mycorrhizal infection can increase phosphate uptake by enabling them to explore more soil. This is obvious where growth of roots is promoted and absorbing rootlets become branched, processes in which auxins are implicated.[77] The increased diameter of the absorbing rootlets due to the fungal sheath and the radially elongated epidermal cells may also increase the amount of accessible soluble phosphate but only by 10% or less in a 1 mm zone of phosphate depletion around the root—this is based on an area of $1.5 \times \pi$ mm^2 around an infected rootlet with an assumed diameter of $500\,\mu m$ compared with $1.4 \times \pi$ mm^2 around a $400\,\mu m$ thick uninfected rootlet.

The loose hyphae extending out from the sheath exploit soil beyond the P-depletion zone and ramify in the phosphate-rich humus, thereby greatly increasing the area of absorbing surface in contact with the labile phosphate pool. Mycorrhizal hyphae can transport phosphate from an external source (labelled with ^{32}P) back to a host plant, where it is translocated to the shoots.[55] Stone[82] found an extensive development of external mycelium associated with seedling uptake of phosphate which probably entered through the fungal mycelium rather than from labile phosphate released into the soil by the fungus. That the mycorrhizal fungi depleted rather than contributed to the labile pool is evident from the decreased uptake of phosphate by Sudan grass grown with *Pinus radiata* seedlings that were mycorrhizal compared with those that were not; the grass also obtained less phosphate from soil in which mycorrhizal pine seedlings had grown compared with soil in which non-mycorrhizal seedlings had grown.

Physiological factors include avidity for and accumulation of phosphate, root longevity and enzyme production. Harley and McCready[41] dissected beech mycorrhizas and showed that the fungal sheath had a greater avidity for phosphate than the host root. This would be a great advantage in the keen competition for phosphate by soil organisms. The sheath can accumulate phosphate which is later released to the host plant in conditions of phosphate deprivation, a useful ecological attribute because flushes of phosphate occur in natural conditions. Roots function longer if they are mycorrhizal[17] and so could continue to use the small amounts of phosphate diffusing very slowly into the P-depletion zone. Considerable phosphatase activity has been demonstrated with excised beech mycorrhizas,[11] and some mycorrhizal fungi produce much phytase in culture.[88] Whether such enzyme action occurs on a large enough scale in nature markedly to increase the pool of labile phosphate requires further investigation. Also, mycorrhizal species differ in their ability to use phosphates of differing availability.[54]

Vesicular–arbuscular mycorrhiza

The vesicular–arbuscular (VA) mycorrhizal fungi are species of *Endogone* and infect plants of most families so far examined, including those of greatest agricultural importance—the Gramineae and Leguminosae.[30, 40, 59, 65] In many low-phosphate soils (see *Table 4.2*) inoculation of plants with *Endogone* increases phosphate uptake by several times, comparable to the effect of adding much monocalcium phosphate.[43, 59] An obvious interpretation is that the mycorrhizal plants can use forms of soil phosphate that non-mycorrhizal plants cannot, thereby short-circuiting the phosphorus cycle. Indeed earlier experiments[25, 63] showed that plants benefit from VA mycorrhiza in the presence of relatively insoluble phosphates such as bone meal and rock phosphate, which suggests solubilisation by the VA fungi.

The only way to obtain direct evidence on solubilisation is to label the labile P in a range of soils and examine the ratio of ^{32}P to ^{31}P

taken up by plants with and without VA mycorrhiza. This ratio would be lower in the mycorrhizal plants if they could obtain some P from insoluble, more weakly labelled, sources. Accordingly, Hayman and Mosse[44] grew onions with and without mycorrhiza in eight different ^{32}P-labelled soils. They found, however, that although the amounts of P absorbed by the plants varied greatly in different soils, the ^{32}P/^{31}P ratio for mycorrhizal and non-mycorrhizal plants in any one soil was always the same and matched that in the soil solution. These results indicate a more efficient absorption of labile P by mycorrhizal plants and not solubilisation. Ross and Gilliam,[71] who determined yields of soybeans in soils supplied with different forms of phosphate and measured the depletion of different phosphate fractions, reached a similar conclusion—namely, that the principal form of phosphate utilised by mycorrhizal plants is the one most readily available to plants irrespective of whether they have mycorrhiza. Hayman and Mosse attributed their own results to the ability of the mycorrhizal hyphae to extend beyond the P-depletion zone to where they could tap the undiluted soluble soil phosphate.

The soil that produced the largest difference in %P between mycorrhizal and non-mycorrhizal plants[44] was examined in more detail by Sanders and Tinker.[74] They, too, obtained the same specific activity of P in mycorrhizal plants, non-mycorrhizal plants and the soil solution. They calculated the theoretical maximum inflow by diffusion to be 3×10^{-14} mol (cm root)$^{-1}$ s^{-1}, which corresponded to the measured figures of 1.6–6×10^{-14} for the non-mycorrhizal plants but not to the figures of 13–22×10^{-14} for the mycorrhizal plants. They later[75] weighed the *Endogone* mycelium formed around the roots and calculated it to be sufficient to account for the observed inflow difference between mycorrhizal and non-mycorrhizal roots. On the basis of counts of entry points and diameters of hyphae, they considered most of the calculated phosphate flux of 3.8×10^{-8} mol cm^2 s^{-1} to be probably due to mass (cytoplasmic) flow through the hyphae, but this needs to be measured experimentally. Bieleski[14] calculated that one millimetre of mycorrhizal root connected to the soil by four hyphae each 25 µm in diameter and 20 mm long would have a P uptake 60 times greater than the same length of uninfected root if diffusion were limiting and 10 times greater if uptake were proportional to surface area. It has recently been shown[42] that VA mycorrhizal hyphae can transport ^{32}P across at least 15 mm of soil and into an infected root. This is direct evidence that the mycelial network around VA mycorrhizal roots can extend the region of phosphate removal well beyond the P-depletion zone.

The hyphal strands growing out into the soil from mycorrhizal roots are somewhat analogous to root hairs, and Baylis[12] suggested that the virtual lack of root hairs in such plants as *Coprosma robusta* could explain their great dependence on mycorrhiza for P uptake in P-deficient soils unless given soluble phosphate. Other plants with poor root hair development, such as onions[43] and citrus,[48] also benefit greatly from VA mycorrhiza. The poor growth of citrus seedlings in fumigated nursery soils in California and Florida was attributed to the elimination of mycorrhizal propagules by the fumigant;[48] the stunted plants grew normally after inoculation with *Endogone*.

Some plants with extensive root hair development do, however, benefit from VA mycorrhiza, e.g. *Paspalum notatum*,[62] tomato[65] and wheat.[47] This suggests that more is involved than just the physical effect of the hyphae exploiting more soil. VA mycorrhizas can indeed accumulate more P than uninfected roots.[32] When an attempt was made[62] to provide the most likely conditions for solubilisation to occur, i.e. in soils (^{32}P-labelled) extremely deficient in phosphate, two somewhat unexpected results were obtained. Firstly, mycorrhizal and non-mycorrhizal plants of *Melinis minutiflora* took up P with the same specific activity, indicating no solubilisation even in conditions of extreme phosphate stress. Secondly, *Paspalum notatum* and *Centrosema pubescens* took up no P at all unless they were mycorrhizal. This latter result suggested either that VA mycorrhizal roots or *Endogone* hyphae have no threshold value below which they cannot take up soil phosphate or that it is lower than that for non-mycorrhizal roots. This potential ability to exhaust the soil solution to below concentrations available to the root suggests a big advantage to mycorrhizal plants growing in conditions in which roots are so close together, as in woodlands and pastures, that depletion zones overlap. The increased avidity of mycorrhizal hyphae for soluble phosphate would permit of better uptake in competition with other fungal hyphae and with bacteria. Differences in threshold values for phosphate uptake between different plants would be another factor making some plants more dependent than others on mycorrhiza.

4.8 Conclusions

Since plants never grow in cultivated or uncultivated soils without a rich population of micro-organisms around their roots, it is unrealistic to assess plant uptake of soil P without considering the activities of these microbes. Harmful microbes are discussed in other texts, and include root pathogens,[7] root inhibitors[17] and those which compete with roots for essential nutrients. Beneficial ones include those which produce growth-promoting substances, bring orthophosphate into the soil solution or directly transport phosphate into plant roots.

Clearly, many bacteria and fungi are capable of dissolving complex inorganic and organic soil phosphates when conditions are right. Two processes must take place in soil, because complex molecules such as phytates must first be brought into solution before orthophosphate ions can be detached by hydrolysis. The solubilisation of complex inorganic phosphates ('microbial weathering') can only occur at localised sites where there is enough energy substrate. The death and autolysis of a temporarily increased microbial biomass will release some orthophosphate into the soil solution. Whether these processes happen at enough microsites and on a large enough scale to benefit the plant is debatable, especially as other organisms and chemical adsorption sites are also vying for this phosphate. The small size of bacterial cells and fungal hyphae, however, permits of closer contact than by roots with the surfaces of insoluble phosphates and, hence, a more efficient mopping up of ions dissociating at the surfaces of these particles.

Studies to elucidate whether the increased growth of plants, which sometimes results from inoculation of seeds with phosphate-solubilising or mineralising bacteria, is due to any extra phosphate produced or to the production of growth-stimulating substances deserve encouragement. They must also verify that populations of selected bacteria coated on the seeds are maintained in the developing rhizosphere. The search for 'superstrains' with both good dissolving ability and good survival qualities should continue. The pelleting of seeds with substrates such as rock phosphate along with the bacteria might be feasible.

Symbiotic micro-organisms have an enormous advantage over free-living micro-organisms because of their niche in plant roots. This makes mycorrhizal fungi much better placed than phosphate-dissolving fungi and bacteria to increase P uptake by plant roots. Their role in P cycling in natural ecosystems is vital. In crop production mycorrhizal inoculation of plants such as citrus seedlings, which are especially dependent on mycorrhiza for nutrient uptake, may well be a practical alternative to large applications of fertiliser, especially where fumigants have been applied to eliminate plant pathogens. The successful inoculation of pine seedlings in countries where pines—and, hence, their mycorrhizal symbionts—are not indigenous is well known. Specificity in the amount of host stimulation[58] makes it necessary to choose the optimum host–fungal strain combination for a particular soil. Even though the VA mycorrhizal fungi are ubiquitous, a selected strain may be much more beneficial than the indigenous ones.[60] Until *Endogone* is grown in pure culture, inoculum could be produced in quantity as spores and infected roots from plants inoculated in small fumigated field plots.

The greater depletion of soluble phosphate by mycorrhizal plants is not of long-term benefit in agriculture. There is therefore a need to study the effects of mycorrhiza on the ability of crop plants, especially the poor phosphate feeders, to utilise less expensive forms of phosphate such as rock phosphate applied to soil and also the residual fertiliser phosphate. The importance of the *Rhizobium* symbiosis in legume production is well known. It may now be rivalled by the importance of the mycorrhizal symbiosis on a broader scale, because most soils in the world are phosphate-deficient and the mycorrhizal condition is the norm for most plant species so far examined.

REFERENCES

1. ADAMS, F. and CONRAD, J. P. Transition of phosphite to phosphate in soils, *Soil Science*, **75**, 361–371 (1953)
2. AGNIHOTRI, V. P. Solubilisation of insoluble phosphates by some soil fungi isolated from nursery seedbeds, *Canadian Journal of Microbiology*, **16**, 877–880 (1970)
3. ALEXANDER, M. *Introduction to Soil Microbiology*, 353–369. Wiley, New York (1961)
4. ANDERSON, G. Nucleic acids, derivatives and organic phosphates, in *Soil Biochemistry*, Volume 1 (edited by A. D. McLaren and G. H. Peterson), 67–90. Arnold, London (1967)
5. AZCÓN, R., BAREA, J. M. and CALLAO, V. Inoculación conjunta de microorganismos movilizadores de fósforo y *Rhizobium* en cultivos enarenados de judia. I. Efectos sobre la parte aérea de las plantas en experiencia hasta floración, *Microbiología Española*, **26**, 31–39 (1973)
6. BAGSHAW, R., VAIDYANATHAN, L. V. and NYE, P. H. The supply of nutrient ions by diffusion to plant roots in soil. V. Direct determination of labile phosphate concentration gradients in a sandy soil induced by plant uptake, *Plant and Soil*, **37**, 617–626 (1972)
7. BAKER, K. F. and SNYDER, W. C. (editors). *Ecology of Soil-borne Plant Pathogens*. University of California Press, Berkeley (1965)
8. BARBER, D. A. Effect of micro-organisms on nutrient absorption by plants, *Nature*, **212**, 638–640 (1966)

9. BARBER, D. A. Micro-organisms and the inorganic nutrition of higher plants, *Annual Review of Plant Physiology*, **19**, 71–88 (1968)

10. BAREA, J. M., RAMOS, A. and CALLAO, V. Contribución al estudio *in vitro* de la mineralización bacteriana de fosfatos, *Microbiología Española*, **23**, 257–270 (1970)

11. BARTLETT, E. M. and LEWIS, D. H. Surface phosphatase activity of mycorrhizal roots of beech, *Soil Biology and Biochemistry*, **5**, 249–257 (1973)

12. BAYLIS, G. T. S. Root hairs and phycomycetous mycorrhizas in phosphorus-deficient soil, *Plant and Soil*, **33**, 713–716 (1970)

13. BHAT, K. K. S. and NYE, P. H. Diffusion of phosphate to plant roots in soil. I. Quantitative autoradiography of the depletion zone, *Plant and Soil*, **38**, 161–175 (1973)

14. BIELESKI, R. L. Phosphate pools, phosphate transport, and phosphate availability, *Annual Review of Plant Physiology*, **24**, 225–252 (1973)

15. BLACK, C. A. *Soil Plant Relationships*. Wiley, New York (1968)

16. BOWEN, G. D. and ROVIRA, A. D. Microbial factor in short-term phosphate uptake studies with plant roots, *Nature*, **211**, 665–666 (1966)

17. BOWEN, G. D. and ROVIRA, A. D. The influence of micro-organisms on growth and metabolism of plant roots, in *Root Growth:* Proceedings of 15th Easter School in Agricultural Science, University of Nottingham (edited by W. J. Whittington). Butterworths, London (1969)

18. BROWN, M. E. Seed and root bacterisation, *Annual Review of Phytopathology*, **12**, 311–331 (1974)

19. BUCKMAN, H. O. and BRADY, N. C. *The Nature and Properties of Soils*. Macmillan, London (1969)

20. CHANG, S. C. and JACKSON, M. L. Fractionation of soil phosporus, *Soil Science*, **84**, 133–144 (1957)

21. COLE, D. W., GESSEL, S. P. and DICE, S. F. Distribution and cycling of nitrogen, phosphorus, potassium and calcium in a second-growth Douglas-fir ecosystem, in *Symposium on Primary Producitivity and Mineral Cycling in Natural Ecosystems* (edited by H. E. Young). University of Maine Press, Orono (Ecological Society of America) (1967)

22. COOKE, G. W. *The Control of Soil Fertility*. Crosby Lockwood, London (1967)

23. COOPER, R. Bacterial fertilizers in the Soviet Union, *Soils and Fertilizers*, **22**, 327–333 (1959)

24. COSGROVE, D. J. Metabolism of organic phosphates in soil, in *Soil Biochemistry*, Volume 1 (edited by A. D. McLaren and G. H. Peterson). Edward Arnold, London (1967)

25. DAFT, M. J. and NICOLSON, T. H. Effect of *Endogone* mycorrhiza on plant growth, *New Phytologist*, **65**, 343–350 (1966)

26. DUFF, R. B., WEBLEY, D. M. and SCOTT, R. O. Solubilisation of minerals and related materials by 2-ketogluconic acid-producing bacteria, *Soil Science*, **95**, 105–114 (1963)

27. DUVIGNEAUD, P. and DENAEYER-DESMET, A. Biological cycling of minerals in temperate deciduous forests, in *Analysis of Temperate Forest Ecosystems*. (edited by D. E. Reichle). Springer-Verlag, Heidelberg (1970)

28. ESTERMANN, EVA and McLAREN, A. D. Contribution of rhizoplane organisms to the total capacity of plants to utilize organic nutrients, *Plant and Soil*, **15**, 243–260 (1961)

29. FARR, E and VAIDYANATHAN, L. V. The supply of nutrient ions by diffusion to plant roots in soil. IV. Direct measurement of changes in labile phosphate content in soil near absorbing roots, *Plant and Soil*, **37**, 609–616 (1972)

30. GERDEMANN, J. W. Vesicular–arbuscular mycorrhiza and plant growth, *Annual Review of Phytopathology,* **6,** 397–418 (1968)

31. GERRETSEN, F. C. The influence of micro-organisms on the phosphate intake by the plant, *Plant and Soil,* **1,** 51–81 (1948)

32. GRAY, L. E. and GERDEMANN, J. W. Uptake of phosphorus-32 by vesicular–arbuscular mycorrhizae, *Plant and Soil,* **30,** 415–422 (1969)

33. GREAVES, S. R., ANDERSON, G. and WEBLEY, D. M. A rapid method of determining the phytase activity of soil micro-organisms, *Nature,* **200,** 1231–1232 (1963)

34. GREAVES, M. P., ANDERSON, G. and WEBLEY, D. M. The hydrolysis of inositol phosphates by *Aerobacter aerogenes, Biochimica Biophysica Acta,* **132,** 412–418 (1967)

35. GREAVES, M. P., VAUGHAN, D. and WEBLEY, D. M. The degradation of nucleic acids by *Cytophaga johnsonii, Journal of Applied Bacteriology,* **33,** 380–389 (1970)

36. GREAVES, M. P. and WEBLEY, D. M. A study of the breakdown of organic phosphates by micro-organisms from the root region of certain pasture grasses, *Journal of Applied Bacteriology,* **28,** 454–465 (1965)

37. GREAVES, M. P. and WEBLEY, D. M. The hydrolysis of *myo*-inositol hexaphosphate by soil micro-organisms, *Soil Biology and Biochemistry,* **1,** 37–43 (1969)

38. GREAVES, M. P. and WILSON, M. J. The degradation of nucleic acids and montmorillonite–nucleic-acid complexes by soil micro-organisms, *Soil Biology and Biochemistry,* **2,** 257–268 (1970)

39. HALM, B. J., STEWART, J. W. B. and HALSTEAD, R. L. The phosphorus cycle in a native grassland ecosystem, in *Isotopes and Radiation in Soil–Plant Relationships including Forestry. International Atomic Energy Agency, Vienna* (1972)

40. HARLEY, J. L. *The Biology of Mycorrhiza,* Leonard Hill, London (1969)

41. HARLEY, J. L. and McCREADY, C. C. The uptake of phosphate by excised mycorrhizal roots of the beech. III. The effect of the fungal sheath of the availability of phosphorus to the core, *New Phytologist,* **51,** 342–348 (1952)

42. HATTINGH, M. J., GRAY, L. E. and GERDEMANN, J. W. Uptake and translocation of ^{32}P-labelled phosphate to onion roots by endomycorrhizal fungi, *Soil Science,* **116,** 383–387 (1973)

43. HAYMAN, D. S. and MOSSE, B. Plant growth responses to vesicular–arbuscular mycorrhiza. I. Growth of *Endogone*-inoculated plants in phosphate-deficient soils, *New Phytologist,* **70,** 19–27 (1971)

44. HAYMAN, D. S. and MOSSE, B. Plant growth responses to vesicular–arbuscular mycorrhiza. III. Increased uptake of labile P from soil, *New Phytologist,* **71,** 41–47 (1972)

45. JOHNSTON, W. B. and OLSEN, R. A. Dissolution of fluorapatite by plant roots, *Soil Science,* **114,** 29–36 (1972)

46. KATZNELSON, H. and BOSE, B. Metabolic activity and phosphate-dissolving capability of bacterial isolates from wheat roots, rhizosphere, and non-rhizosphere soil, *Canadian Journal of Microbiology,* **5,** 79–85 (1959)

47. KHAN, A. G. The effect of vesicular–arbuscular mycorrhizal associations on growth of cereals. II. Effects on wheat growth, *Annals of Applied Biology* (in press) (1975)

48. KLEINSCHMIDT, G. D. and GERDEMANN, J. W. Stunting of citrus seedlings in fumigated nursery soils related to the absence of endomycorrhizae, *Phytopathology,* **62,** 1447–1453 (1972)

49. KOBUS, J. The distribution of micro-organisms mobilising phosphorus in different soils, *Acta Microbiologica Polonica,* **11,** 255–264 (1962)

50. LARSEN, S. Soil phosphorus, *Advances in Agronomy,* **19,** 151–210 (1967)

51. LEWIS, D. G. and QUIRK, J. P. Phosphate diffusion in soil and uptake by plants. III. P^{31} movement and uptake by plants as indicated by P^{32}-autoradiography, *Plant and Soil,* **26,** 445–453 (1967)

52. LOUW, H. A. and WEBLEY, D. M. A plate method for estimating the numbers of phosphate dissolving and acid producing bacteria in soil, *Nature*, **182**, 1317–1318 (1958)

53. MARTIN, J. K. The influence of rhizosphere microflora on the availability of ^{32}P-*myo*-inositol hexaphosphate phosphorus to wheat, *Soil Biology and Biochemistry*, **5**, 473–483 (1973)

54. MEJSTRIK, V. K. and KRAUSE, H. H. Uptake of ^{32}P by *Pinus radiata* roots inoculated with *Suillus luteus* and *Cenococcum graniforme* from different sources of available phosphate, *New Phytologist*, **72**, 137–140 (1973)

55. MELIN, E. and NILSSON, H. Transfer of radioactive phosphorus to pine seedlings by means of mycorrhizal hyphae, *Physiologia Plantarum*, **3**, 88–92 (1950)

56. MISHUSTIN, E. N. Processus microbiologiques mobilisant les composés du phosphore dans le sol, *Revue d'Écologie et de Biologie du Sol*, **9**, 521–528 (1972)

57. MISHUSTIN, E. M. and NAUMOVA, A. M. Bacterial fertilizers, their effectiveness and mode of action, *Microbiology* [Translations of *Mikrobiologiya*], **31**, 442–452 (1962)

58. MOSSE, B. The influence of soil type and *Endogone* strain on the growth of mycorrhizal plants in phosphate deficient soils, *Revue d'Écologie et de Biologie du Sol*, **9**, 529–537 (1972)

59. MOSSE, B. Advances in the study of vesicular–arbuscular mycorrhiza, *Annual Review of Phytopathology*, **11**, 171–196 (1973)

60. MOSSE, B. and HAYMAN, D. S. Plant growth responses to vesicular–arbuscular mycorrhiza. II. In unsterilized field soils, *New Phytologist*, **70**, 29–34 (1971)

61. MOSSE, B. and HAYMAN, D. S. Mycorrhiza in Meathop Wood, in *Ecosystem Study of Meathop Wood*. (in press) (1974)

62. MOSSE, B., HAYMAN, D. S. and ARNOLD, D. J. Plant growth responses to vesicular–arbuscular mycorrhiza. V. Phosphate uptake by three plant species from P-deficient soils labelled with ^{32}P, *New Phytologist*, **72**, 809–815 (1973)

63. MURDOCH, C. L., JACKOBS, J. A. and GERDEMANN, J. W. Utilization of phosphorus sources of different availability by mycorrhizal and non-mycorrhizal maize, *Plant and Soil*, **27**, 329–334 (1967)

64. NAGARAJAH, S., POSNER, A. M. and QUIRK, J. P. Competitive adsorption of phosphate with polygalacturonate and other organic anions on kaolinite and oxide surfaces, *Nature*, **228**, 83–85 (1970)

65. NICOLSON, T. H. Vesicular–arbuscular mycorrhiza—a universal plant symbiosis, *Science Progress, Oxford*, **55**, 561–581 (1967)

66. NYE, P. H. and FOSTER, W. N. M. A study of the mechanism of soil-phosphate uptake in relation to plant species, *Plant and Soil*, **9**, 338–352 (1958)

67. PROBERT, M. E. The dependence of isotopically exchangeable phosphate (L-value) on phosphate uptake, *Plant and Soil*, **36**, 141–148 (1972)

68. RIDGE, E. H. and ROVIRA, A. D. Phosphatase activity of intact young wheat roots under sterile and non-sterile conditions, *New Phytologist*, **70**, 1017–1026 (1971)

69. ROBERTSON, R. A. and DAVIES, G. E. Quantities of plant nutrients in heather ecosystems, *Journal of Applied Ecology*, **2**, 211–219 (1965)

70. RORISON, I. H. The response to phosphorus of some ecologically distinct plant species. I. Growth rates and phosphorus absorption, *New Phytologist*, **67**, 913–923 (1968)

71. ROSS, J. P. and GILLIAM, J. W. Effect of *Endogone* mycorrhiza on phosphorus uptake by soybeans from inorganic phosphates, *Soil Science Society of America Proceedings*, **37**, 237–239 (1973)

72. RUSSELL, E. W. *Soil Conditions and Plant Growth*, 10th edition. Longman, London (1973)

73. SACKETT, W. G., PATTEN, A. J. and BROWN, C. W. The solvent action of soil bacteria upon the insoluble phosphates of raw bonemeal and natural raw rock phosphate, *Zentralblatt für Bakteriologie, Parasitenkunde, und Infektionskrankheiten (IIte Abteilung)*, **20**, 688–703 (1908)

74. SANDERS, F. E. and TINKER, P. B. Mechanism of absorption of phosphate from soil by *Endogone* mycorrhizas, *Nature*, **233**, 278–279 (1971)

75. SANDERS, F. E. and TINKER, P. B. Phosphate flow into mycorrhizal roots, *Pesticide Science*, **4**, 385–395 (1973)

76. SAUCHELLI, V. *Phosphates in Agriculture*. Reinhold, New York; Chapman and Hall, London (1965)

77. SLANKIS, V. Formation of ectomycorrhizae of forest trees in relation to light, carbohydrates and auxins, in *Mycorrhizae*. Proceedings of the first North American Conference on Mycorrhizae, April 1969: Miscellaneous Publication 1189, US Department of Agriculture—Forest Service (1971)

78. SMITH, J. H., ALLISON, F. E. and SOULIDES, D. A. Evaluation of phosphobacterin as a soil inoculant, *Soil Science Society of America Proceedings*, **25**, 109–111 (1961)

79. SPERBER, J. I. Solution of mineral phosphates by soil bacteria, *Nature*, **180**, 994–995 (1957)

80. SPERBER, J. I. The incidence of apatite-solubilizing organisms in the rhizosphere and soil, *Australian Journal of Agricultural Research*, **9**, 778–787 (1958)

81. STARK, N. Nutrient cycling. I. Nutrient distribution in some Amazonian soils, *Tropical Ecology*, **12**, 24–50 (1971)

82. STONE, E. L. Some effects of mycorrhizae on the phosphorus nutrition of Monterey pine seedlings, *Soil Science Society of America Proceedings*, **14**, 340–345 (1949)

83. SUTTON, C. D. and GUNARY, D. Phosphate equilibria in soil, in *Ecological Aspects of the Mineral Nutrition of Plants* (edited by I. H. Rorison). Blackwell, Oxford (1969)

84. SWABY, R. J. and SPERBER, J. Phosphate dissolving micro-organisms in the rhizosphere of legumes, in *Nutrition of 'Legumes* (edited by E. D. Hallsworth), 289–294. Butterworths, London (1959)

85. SWITZER, G. L. and NELSON, L. E. Nutrient accumulation and cycling in Loblolly Pine (*Pinus taeda* L.) plantation ecosystems: the first twenty years, *Soil Science Society of America Proceedings*, **36**, 143–147 (1972)

86. TARDIEUX-ROCHE, A. Contribution à l'étude des interactions entre phosphates naturels et microflore du sol, *Annales Agronomiques*, **17**, 403–479 (1966)

87. TARDIEUX-ROCHE, A. and TARDIEUX, P. La biosynthèse des phosphates condensés par la microflore du sol et son rôle dans la nutrition des végétaux, *Annales Agronomiques*, **21**, 305–314 (1970)

88. THEODOROU, C. Inositol phosphates in needles of *Pinus radiata* D. Don and the phytase activity of mycorrhizal fungi, *Transactions of the 9th International Congress of Soil Science*. Volume III, 483–490 (1968)

89. TSUBOTA, G. Phosphate reduction in the paddy field I, *Soil and Plant Food*, **5**, 10–15 (1959)

90. WILD, A. and OKE, O. L. Organic phosphate compounds in calcium chloride extracts of soils: identification and availability to plants, *Journal of Soil Science*, **17**, 356–371 (1966)

91. WOOLHOUSE, H. W. Differences in the properties of the acid phosphatases of plant roots and their significance in the evolution of edaphic types, in *Ecological Aspects of the Mineral Nutrition of Plants*. (edited by I. H. Rorison), Blackwell, Oxford (1969)

5 Extracellular Polysaccharides of Soil Bacteria

CHRISTINE M. HEPPER

Bacteriologists are familiar with the capsules surrounding many bacteria. These capsules, which can be observed by negative staining with Indian ink or Nigrosine,[20] form a distinct layer into which the particles of ink or Nigrosine penetrate only slowly (*Figure 5.1*). Many

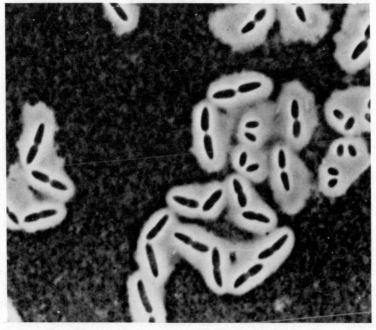

Figure 5.1 Cells of Azotobacter *sp. dispersed in Indian ink*

bacteria also synthesise extracellular polysaccharides which diffuse away from the cells into the liquid medium or form a mucoid layer around organisms growing on an agar surface. Deinema and Zevenhuisen[19] observed that cellulose fibrils are commonly found in the extracellular polysaccharides of bacteria which form flocs in liquid culture, including species of *Azotobacter, Pseudomonas, Rhizobium* and *Agrobacterium.* They suggested that these fibrils help to retain the polysaccharide around groups of cells, and that when digested with cellulase, the flocs are easily dispersed (*Figure 5.2*). Generally, the capsular and extracellular polysaccharides have the same composition, and it has been suggested that the difference between bacteria synthesising soluble polysaccharide and those producing only a capsule is either the absence of an enzyme which binds the polysaccharide to the cell wall or the loss of specific sites on the cell wall where the polysaccharide would be attached.[66] The biosynthesis

Figure 5.2 Cells of Pseudomonas *sp. showing a background of cellulose fibrils (reproduced by permission of Springer-Verlag, Heidelberg)*

of these polysaccharides has been studied and, in the case of those composed of more than one sugar, is found to require several specific enzymes. These effect the formation and interconversion of sugar nucleotides, the transfer of sugar moities to a lipid carrier and subsequent polymerisation.[66]

Despite much work on capsules and extracellular polysaccharides, their functions remain obscure. Since non-mucoid mutants of many bacteria can grow and reproduce quite well in laboratory cultures, these polysaccharides cannot be essential components. Capsules can also be removed enzymically from some bacteria without loss of viability. Various hypotheses have been put forward to explain the functions of capsular and extracellular polysaccharides and their value to the particular organism in its normal environment.

These polysaccharides are highly hydrated and may protect the cells during conditions of water stress. Under such conditions, water present in the polysaccharide matrix could be utilised by the bacteria. The polysaccharides are also hygroscopic and may allow the cells to rehydrate slowly after desiccation.[72] It has also been suggested that the polysaccharide might protect the bacteria from attack by protozoa or bacteriophages. This latter idea seems unlikely, however, since in many systems involving bacteria and bacteriophages a specific depolymerase is produced[65] which breaks down the polysaccharide into low molecular weight repeating-units, thus solubilising it and presumably facilitating attachment of the bacteriophages to the exposed cell walls. Postgate[53] proposed a novel function for polysaccharides in capsular bacteria which are free-living nitrogen fixers. He suggested that a thick layer of slime around colonies of *Derxia gummosa* would diminish diffusion of oxygen to the cells, so providing conditions favourable for nitrogen fixation. Barooah and Sen[8] found a correlation between nitrogen fixation and polysaccharide production in *Beijerinckia,* but a strain of *Azotobacter vinelandii,* unable to synthesise polysaccharide, did fix nitrogen at normal oxygen concentrations, showing that this protection was not obligatory.[10] In at least one case, bacterial extracellular polysaccharide has been shown to act as a reserve carbon source for the organisms. Cultures of *Beijerinckia indica* synthesised polysaccharide, giving a maximum yield after 9 days' growth; then, when the glucose was depleted, continued to grow at the expense of their extracellular polysaccharide material.[43] This organism was, however, unable to utilise the polysaccharide as a carbon source after it had been isolated and purified, which showed, perhaps, that some changes had occurred during purification.

Extracellular polysaccharides synthesised by the bacteria have been implicated in the infection of legumes by *Rhizobium*.[42] It was suggested that the formation of polygalacturonase by the potential host was induced by the specific bacterial polysaccharides. This enzyme then facilitated entry of the rhizobia into the plant root hairs, which is essential to the establishment of the symbiosis. There has been no confirmatory evidence for this suggestion but it remains possible that the polysaccharide is the specific determinant in the symbiosis—that is, the means by which the host 'recognises' the bacterium—since polysaccharide–protein interactions are known to be very specific, depending often on minor sugar constituents.

The composition of capsules and extracellular polysaccharides synthesised by bacteria in culture has received considerable attention and recently more work has been done on the structure of these polymers. The most important condition for studying these polysaccharides is to be absolutely sure that they are homogeneous. Purity should be checked by electrophoresis, sedimentation analysis or column chromatography; and unless this is done, subsequent analyses or structural studies may be inaccurate. The identification of sugars has been greatly improved by the use of paper chromatography, although this does not always give adequate separation of some sugars commonly found in bacterial polysaccharides which differ only in the configuration at one carbon atom. Better methods for both quantitative and qualitative analysis are gas–liquid chromatography of volatile sugar derivatives and the use of specific enzyme assays which will also distinguish between different optical isomers of a sugar. Structural studies have also been facilitated by the use of phage-induced depolymerases.[67] It is possible that workers not using these methods may have missed minor components in polysaccharides and often assays for non-sugar substituents were not carried out. One must bear in mind these limitations when considering the results of assays of extracellular polysaccharides.

Soil bacteria grown in pure culture produce either heteropolysaccharides—that is, polymers composed of several types of monosaccharide together with, in some cases, non-sugar substituents—or homopolysaccharides, which consist of only one type of sugar residue, usually linked together in the same way. Either type may be neutral polymers or be composed of sugar acids and, in addition, other organic acids. Pyruvic acid has recently been found to be a component of some bacterial polysaccharides and it may be more common than had previously been thought. In the ex-

96

tracellular polysaccharide obtained from *Xanthomonas campestris* pyruvyl groups are attached terminally through the ketone carbon to the hydroxyl groups of the fourth and sixth carbon atoms of neutral hexose residues, forming additional six-membered heterocyclic rings,[60] and this type of structure may well be common to other pyruvate-containing polysaccharides. Zevenhuizen[74] found evidence that pyruvate could also be attached to glucose molecules present in a (1→3)-linked chain. *O*-Acetyl groups have also been commonly found attached to free hydroxyls of polysaccharides.

A typical example of an acidic heteropolysaccharide is that synthesised by *Arthrobacter viscosus,* which is composed of repeating units made up of equal proportions of D-mannuronic acid, D-glucose and D-galactose, the whole polymer being 25% acetylated.[38]

Several types of soil bacteria elaborate polysaccharides which contain amino sugars, sometimes together with uronic acids. The polysaccharide obtained from *Chromobacterium violaceum* is composed of glucose, probably glucuronic acid and an amino sugar in the molar ratio 5 : 1 : 1;[17] and another species which synthesises an acidic capsular polysaccharide containing several amino sugars is *Achromobacter georgiopolitanum.* This capsular polysaccharide is composed of 2-acetamido-2-deoxy-D-glucuronic acid, 2-acetamido-2-deoxy-glucosamine and 2-acetamido-2,6-dideoxy-D-glucosamine in a 2:1:2 molar ratio; another non-acidic fucosamine (2-amino-2,6-dideoxy-D-galactose)-rich polysaccharide is also synthesised by this bacterium.[61]

The unpurified extracellular material obtained from *Serratia marcescens* was thought to be a mixture of polysaccharides, of which two were probably loosely linked together. One of these contained rhamnose and glucose in the main chain; and the other, glucose, L-glycero-D-mannoheptose and D-glycero-D-mannoheptose.[73] These polysaccharides contained smaller amounts of galactose, galactosamine, glucosamine and bound lipid. Another acidic polysaccharide was isolated from cultures of this bacterium and found to consist of D-glucose and D-mannose with a small amount of D-glucuronic acid joined together in a main chain through (1→3)-linkages. Myristic and hydroxy-myristic acid were also bound to this polymer.[4] Three similar polysaccharides together with a rhamnoheptoglucan were found in the capsule of *Serratia marcescens,* confirming the relationship between the extracellular and capsular polysaccharides.[3]

Four species of *Agrobacterium* (*tumefaciens, rhizogenes, radiobacter* and *rubi*) synthesised homopolysaccharides composed of 1,2-Dglucopyranose units linked with a β-configuration to give a linear molecule. There was evidence that these polymers may have contained unidentified constituents.[27] Other strains of *A. tumefaciens* produced an acidic heteropolysaccharide containing glucose, galactose, pyruvate and acetyl in the molar ratio of 6 : 1 : 1 : 1.[74]

Rhizobium is another genus in which different strains synthesise distinct polysaccharides. The species of *Rhizobium* are designated according to the legume host which they nodulate. *R. japonicum* has been shown to synthesise a neutral polysaccharide which appears to be identical with that produced by many species of *Agrobacterium*.[18] Another link between these two genera is that *R. meliloti* forms a polysaccharide composed of glucose, galactose, pyruvate and acetyl which appears identical in structure with that synthesised by two strains of *A. tumefaciens*.[74] Björndal *et al*.[9] and Zevenhuizen, [74] using slightly differing polysaccharides from various strains of *R. meliloti* and *A. tumefaciens,* suggested similar structures—that is, a main chain composed of both β-$(1\rightarrow 3)$- and β-$(1\rightarrow 4)$-linked sugars. Of the glucose molecules, 12% were branch points linked through the first, fourth and sixth carbon atoms, and the *O*-acetyl groups, also attached to glucose residues, were randomly placed along the chain. Zevenhuisen[74] also found evidence that a small amount of the glucose present was linked through carbon atoms 1 and 6.

R. trifolii, R. phaseoli and *R. leguminosarum* polysaccharides commonly contain glucuronic acid. A typical composition of a polysaccharide synthesised by these bacteria was found by Zevenhuisen[74] to be glucose, galactose, glucuronic acid, pyruvate and *O*-acetyl in the ratio 5 : 1 : 2 : 2 : 3. The structure proposed for this type of polysaccharide was a chain of glucose, galactose and glucuronic acid residues with both $(1\rightarrow 3)$- and $(1\rightarrow 4)$-linkages. Pyruvate substituents were found both along the chain and terminally, attached to glucose or galactose. Some glucose and galactose residues were branch points, being substituted at carbon atoms 1, 4 and 6.[74] In general, the presence or absence of glucuronic acid provides a broad classification of these rhizobial poly-saccharides, but Clapp and Davis[12] studied two strains of *R. meliloti* which had a moderate glucuronic acid content and one strain of *R. trifolii* which had no uronic acid in the extracellular polysaccharide. Various other sugars have been reported in polysaccharides synthesised by *Rhizobium*—for example, 4-*O*-methyl glucuronic acid[37] and glucosamine and galactosamine.[64] It is probable that new components will be identified in polysaccharides when these studies are continued.

The results of analyses of extracellular and capsular polysaccharides from *Azotobacter* species show great diversity both at the specific level and between strains. The extracellular polysaccharide from *A. indicum* was shown to be composed of two polymers: a neutral polysaccharide containing D-glucose, D-mannose, L-arabinose and L-rhamnose (6-deoxy-L-mannose); and the major component (88%)—an acidic linear polymer made up of repeating units composed of equal parts of D-glucuronic acid, D-glycero-D-mannoheptose and D-glucose with α-(1 → 3)- and (1 → 2)-linkages.[26, 51] Haug and Larsen[35] found a similar polysaccharide synthesised by *Beijerinckia indica (Azotobacter indicum)*, but in this case the uronic acid was guluronic acid and some acetyl groups were present. The polysaccharides isolated from *A. chroococcum* cultures contained glucose, galactose, rhamnose and either galacturonic or glucuronic acid,[41, 52] and another strain analysed by Lopez and Becking[43] contained xylose in addition to these sugars. Extracellular polysaccharide isolated from several strains of *A. vinelandii* commonly contained galacturonic acid, glucose and rhamnose with various amounts of *O*-acetyl groups and other sugars.[15, 43, 52] Actual amounts for one particular strain were 38% galacturonic acid, 1.5% glucose, 1.7% rhamnose and 2.9% acetyl.[15] Another strain of *A. vinelandii* produced a very different polysaccharide, composed mainly of D-mannuronic acid with a small proportion of L-guluronic acid units, the whole molecule being partially acetylated.[26] Further work on this type of polysaccharide from various strains of *A. vinelandii* indicated that the structure was very similar to that of alginates obtained from brown algae, and consisted of blocks of D-mannuronic acid units alternating with blocks of L-guluronic acid units with one acetyl group for every five uronic acid residues.[40] The ratio of mannuronic to guluronic acid was about 0.56, but evidence was obtained that this ratio could be modified, after the polymer had been synthesised, by a specific epimerase secreted by the cells.[36] Claus[14] found the unusual sugar, 2 keto-3-deoxygalactonic acid, associated with the extracellular polysaccharide of some strains of *A. vinelandii*. *A. agilis* differs from the other species described in having only capsular polysaccharide, and on analysis this was shown to be composed of galactose and rhamnose in a molar ratio of 1.0 : 0.7 with small amounts of a sialic acid-like component.[16] From the diverse results obtained with different strains of the same *Azotobacter* species it seems that polysaccharide composition is of little value as a classification criterion for this genus.

Other soil bacteria, including *Pseudomonas fluorescens, Alcaligenes*

99

viscosus and several species of *Bacillus (subtilis, megatherium* and *polymyxa)*, all commonly form levans in pure culture.[22, 24, 70] These are polysaccharides which give only fructose on hydrolysis and in the case of *Bacillus* spp. are composed of β-2,6-linked D-fructofuranose units. Some *Bacillus* species synthesise acidic heteropolysaccharides containing glucose, mannose and glucuronic acid in addition to levans.[22] *B. macerans* was unique in synthesising Schardinger dextrins, which are low molecular weight glucans composed of six, seven and eight glucose residues joined into a continuous loop.[23, 59]

A species of *Flavobacterium* was able to synthesise a complex neutral polymer containing D-galactose, D-glucose, D-mannose, D-ribose and L-rhamnose when vanillin and other lignin-derived compounds were used as carbon sources for growth.[54] Polysaccharides isolated from the two plant pathogens, *Corynebacterium insidiosum* and *C. sepedonicum*, were composed of galactose, glucose and fucose (6-deoxygalactose) in the molar ratios of 1.4 : 0.92 : 2 and 1.55 : 0.93 : 1, respectively.[63] In addition, pyruvic acid was found in the polysaccharide from *C. insidiosum*.[25] The fact that these polysaccharides contained fucose, which is not found in healthy plant material, was used to establish that they entered the xylem vessels of plants infected by these bacteria, causing blockage and subsequent wilting.[63]

Only a limited number of bacterial genera are present in soil, but many of these can synthesise polysaccharides. Rouatt and Katznelson[56] found *Arthrobacter* and *Pseudomonas* species to be the most common bacteria in agricultural soils, the numbers of arthrobacters decreasing in the presence of wheat roots and the numbers of pseudomonads increasing. Other common types associated with the wheat rhizosphere were species of *Achromobacter*, *Nocardia* and *Bacillus*. A similar observation on the predominance of *Arthrobacter* species in cultivated and uncultivated soils, where they constitute 30–60% of the bacterial flora, was made by Soumare and Blondeau.[62] These workers also found *Flavobacterium, Achromobacter* and *Pseudomonas* to be common in cultivated soils, whereas *Bacillus* species predominated in fallow soil. Bacteria able to synthesise extracellular polysaccharides in culture, particularly levans, were generally found to be more common on and around the roots of pasture grasses than in non-rhizosphere soil, but those bacteria which had only capsules showed no preference for the root as a natural habitat.[69]

Since polysaccharides are presumably formed in soil and can interact with clays, they have been implicated in the stabilisation of soil aggregates.[44] Martin *et al.*[48] have defined a soil aggregate as 'a

100

naturally occurring cluster or group of soil particles in which the forces holding the particles together are much stronger than the forces between adjacent aggregates.' The type of aggregate formed depends on a balance between the adhesion of particles caused by cementing substances and the fragmentation brought about by mechanical forces such as freezing and root penetration.[48] It is the stabilisation of these aggregates against dispersion by water that is possibly brought about by bacterial polysaccharides. A soil with optimum size aggregates provides good conditions for plant growth, particularly for the penetration and anchorage of roots and the emergence of young shoots. Such a soil has sufficiently large pore spaces to allow of rapid penetration of rainwater but also moderate drainage and water retention so that the soil remains moist but well-aerated. These soils do not form crusts, are resistant to erosion and are generally more suitable for mechanised agricultural practice.[57]

To assess whether bacterial polysaccharides are able to increase soil aggregate stability, three types of experiment have been tried: increasing the carbon content of the soil to stimulate the natural microflora; addition of bacteria to the soil; and incorporation into soil of bacterial polysaccharides isolated from laboratory cultures. No change in aggregation was observed when sterile soils were amended with organic material (see reference 30), but the addition of glucose to soil on a field scale produced a rapid improvement in aggregation, presumably due to microbial activity.[31] This confirms the general observation that addition of plant remains, such as straw, to poorly aggregated soils leads to an improvement in soil structure. The addition of nitrogen together with the organic material gives even better results, presumably because micro-organisms are able to decompose the organic matter completely. Aspiras *et al.*[6] found that the addition of suspensions of four bacteria in sucrose solution improved the binding of soil aggregates. Aggregates stabilised by the addition of suspensions of *Agrobacterium radiobacter* and *Bacillus polymyxa* were superior to those to which *Rhizobium trifolii* and *Agrobacterium tumefaciens* had been added. Swaby,[68] however, found that only a few of the bacteria that he tested were able successfully to stabilise soil aggregates.

It is well established that polysaccharides isolated from pure cultures of bacteria and added to soil do increase the stability of aggregates. Extracellular polysaccharides from *Azotobacter indicum* and *Chromobacterium violaceum* were very effective stabilisers, the latter being superior to any other material tested and comparable with a synthetic soil conditioner.[46, 49] Clapp, Davis and Waugaman[13] and

Harris *et al.*[32] made similar observations on the ability of low concentrations (0.5%) of various rhizobial and other bacterial polysaccharides effectively to stabilise soil particles.

The mechanism of the binding of aggregates by these microbial polysaccharides has been the subject of speculation and study in simplified systems. One method of approach is to equilibrate montmorillonite with polysaccharide solutions of various concentrations and to measure the adsorption. The shape of the isotherms obtained depends on the ionic form of the clay and may be either sigmoid, indicating that at a certain concentration new adsorption sites become available, or Langmuirian, showing that only a fixed number of sites are available, the polysaccharide molecules forming a monolayer on the surface.[21]

Polysaccharides are usually long-chain molecules with, in some cases, side branches, possessing a large number of free hydroxyl groups and, in acidic polysaccharides, free carboxyl groups. The mechanism of binding of polysaccharide molecules to clay particles is thought to be by ionic and hydrogen bonding. The only means of bonding in a levan or dextran would be through hydrogen bonds from unsubstituted hydroxyl groups to the oxygen atoms of the clay. A large number of weak bonds such as these results in the polymer being very difficult to desorb.[28] Hydrogen bonding would also be important in the binding of anionic polysaccharides, with the possibility of additional hydrogen bonds between carboxyl groups and clay particles, as observed between polygalacturonic acid and bentonite.[39] Acidic polysaccharides could not be linked ionically to a negatively charged clay such as montmorillonite, and it has been suggested that in this case calcium might act as a bridging cation.[28] Clapp, Davis and Waugaman[13] have produced evidence which questions the importance of carboxyl groups in the aggregation of soils by rhizobial polysaccharides. They found that high molecular weight polysaccharides with high uronic acid content were poor aggregate stabilisers compared with those with low uronic acid content. From the experiments described it seems that bacterial polysaccharides can stabilise soil aggregates under laboratory conditions, but the question remains as to whether they are the actual agents influencing stability of soils under natural conditions.

Some bacterial polysaccharides are more resistant than plant material to degradation by soil micro-organisms. Martin and Richards[49] found that when a uronic acid-containing polysaccharide synthesised by *Chromobacterium violaceum* was added to soil, approximately half of it remained after 8 weeks. After 2 weeks, 30% of it had disappeared compared with an average of 45% decomposition of added straw and 70% loss of bacterial dextrans and levans. Decomposition of *Chromobacterium* polysaccharide did not begin immediately, which indicated that micro-organisms with suitable degradative enzymes were not prevalent. When a comparison was made of the rates of decomposition of polysaccharides obtained from various soil bacteria, that from *Azotobacter indicum* was the most resistant, 81% of it still remaining after 8 weeks. The figures obtained for other polysaccharides in the same soil were: from *Chromobacterium violaceum,* 61% remaining, from *Azotobacter chroococcum,* 33% remaining; and from *Bacillus polymyxa*—a levan—only 21% remaining. These were all compared with glucose, of which 16% remained. The majority of soil micro-organisms were unable to use these polysaccharides as a carbon source except that from *A. chroococcum* cultures.[46] In addition, Aspiras *et al.*[7] showed that binding material synthesised by micro-organisms in sucrose-amended soil were not resistant to biodegradation.

Although substances which stabilised aggregates were synthesised in sucrose-amended soils under aerobic and anaerobic conditions, when soils became waterlogged, this material was not degraded and the aggregates remained stable for a considerable length of time.[32] When aerobic conditions were restored, rapid decomposition and loss of stability followed. However, in a natural anaerobic soil the presence of polysaccharides caused a reduction in permeability by blocking the soil pores, leading to conditions which could be detrimental to plant growth.[5]

One reason why complex polysaccharides are generally degraded only slowly in soil is that micro-organisms possessing suitable constitutive enzymes are not present, but another consideration is that the polysaccharides are protected from attack by binding with clay material or with soil organic matter. Griffiths and Burns[31] found that polysaccharide from a yeast—*Lipomyces starkeyi*—increased the aggregate stability of some soils and that this polysaccharide could be protected from degradation by infiltrating the soil with tannic acid. A similar result was obtained on a field scale when soil was infiltrated with glucose and tannic acid. The increased stability of the aggregates was maintained for a much longer period of time than when glucose alone was used. Martin, Ervin and Shepherd[47] studied the decomposition in soil of polysaccharides complexed with various heavy metals. The iron, zinc and copper salts of the polysaccharides isolated from *Azotobacter indicum* and *Arthrobacter viscosus* were more resistant to degradation than the aluminium salts or the free acids. A

103

similar result was obtained with the polysaccharides from *A. chroococcum,* except that the zinc complex was readily decomposed. However, all the metal complexes of *Chromobacterium violaceum* polysaccharide were degraded at approximately the same rate and to the same extent as the free acid. The decomposition of all salts and free acids was generally much slower than that of glucose and often showed a lag-phase during which very little material disappeared. Degradation of a fructosan synthesised by *Bacillus subtilis* was only a little slower than the rate of glucose disappearance and was altered very little by the presence of metal ions.

Many workers have tried to improve the stability of a poorly aggregated soil by adding a polysaccharide fraction derived from a stable soil. The material thought most likely to be derived from microbial polysaccharides is that obtained from the fulvic acid fraction by acetone precipitation. Highest estimates of the carbohydrate content of this material were about 50%,[71] the remaining part being predominantly inorganic.[1] Analysis showed a wide range of sugars to be present in this fraction—galactose, mannose, arabinose, rhamnose, glucuronic acid, xylose and glucose. The last two sugars were present in small amounts, which confirmed that the polysaccharides were derived from micro-organisms rather than from plants.[71] The material precipitated from the fulvic acid fraction was shown to have soil-stabilising properties; the extent to which it was effective probably depended on its carbohydrate content.[55, 71] Rennie, Truog and Allen[55] found it less effective than a synthetic soil conditioner or polysaccharide isolated from *Agrobacterium radiobacter* cultures.

It is reasonable to assume that if the presence of microbial polysaccharides is the most important factor in soil aggregation, then there should be a correlation between the observed soil stability and the amount of material precipitated from the fulvic acid fraction. Salomon[58] found that aggregate stability of soils cropped alternately with potatoes and *Agrostis* sp. was not correlated with the polysaccharide or polyuronide content of soil samples taken late in the season. In contrast to these results, Chester, Attoe and Allen,[11] using a different extraction technique, found that the aggregation of a soil was correlated with the microbial polysaccharide content. Acton, Rennie and Paul[2] were only able to demonstrate a 35% correlation between microbial polysaccharide content and the degree of aggregation, and also observed that polysaccharide production reached its maximum after 10 days, whereas the highest levels of soil aggregation occurred after 42 days. For these reasons they considered that aggregation was probably a function of the total carbohydrate content rather than the microbial polysaccharide alone.

Periodic acid readily breaks the bond between two adjacent carbon atoms if both carry a hydroxyl group. When susceptible polysaccharides are treated in this way, the sugar rings are opened and the resulting polyaldehyde is easily degraded under mild alkaline conditions to low molecular weight fragments. As this type of structure is very common in polysaccharides, except where the sugars are predominantly (1→3)-linked, it might be expected that treatment of soil with periodate followed by dilute alkali would destroy aggregates which depended for stability on the presence of polysaccharides. On the basis of periodate treatment of natural and synthetic aggregates, Mehta *et al.*[50] concluded that polysaccharide played no part in the stabilisation of aggregates from a forest soil. Synthetic aggregates prepared by the addition of soil or microbial polysaccharides to the same soil were dispersed by water after similar treatment. Greenland, Lindstrom and Quirk[29] found that soil taken from sites under pasture for several consecutive years was almost completely resistant to treatment by periodate, whereas soils taken from young pastures (2–4 years old) and from intensively cropped fields did depend for their aggregate stability on material which was degraded by periodate. Sub-surface soils from all sites tested lost their stability when subjected to the same treatment. Aspiras *et al.*[6] and Harris *et al.*[32] found that the increase in aggregate stability brought about by the addition of bacterial suspensions in sucrose solution or by bacterial polysaccharides isolated from cultures was lost when the soils were treated with periodate.

The experiments cited indicate that polysaccharides from pure cultures and isolated from well-aggregated soils have the capacity to form artificial, water-stable aggregates. This evidence does not prove that polysaccharides are the only agencies responsible for the stability of soils under natural conditions or indeed whether they are involved at all. From the physical point of view, polysaccharides are ideally suited to the role of soil stabilisers, since these long, flexible molecules, with many positions available for bonding, are able to bridge the spaces between clay particles and form many points of contact with each.[45]

The fact that there is generally poor correlation between aggregate stability and microbial polysaccharide content of a soil is not surprising, since the fraction isolated is impure and does not account for all the microbial polysaccharide in the soil. There are possibilities here for a study of the recovery of bacterial polysaccharides introduced into soils, which might be facilitated by the use of radioactively labelled polymers.

Bacteria which synthesise polysaccharides are more common around plant roots than in the non-rhizosphere soil, and it has been observed that characteristically there are more aggregates present in this region also (see reference 33). This, again, is indirect evidence for the involvement of polysaccharide-synthesising bacteria in the stabilisation of soil around roots. The degree of aggregation of a particular soil may be dependent on the equilibrium established between synthesis and degradation of cementing substances, and this would be related to the prevailing environmental conditions.[7]

Attempted breakdown of aggregates by polysaccharide-oxidising reagents has provided convincing evidence that polysaccharides are the most important substances stabilising soils under annual crops and short-term pastures. Several workers consider that bacterial polysaccharides exert an ephemeral effect and are later replaced by periodate-resistant materials in soils under established vegetation. Harris *et al.*[34] suggested that bacteria may be important in the early stages of soil aggregation by promoting the formation of small aggregates, but later these become stabilised in other ways—for instance, by entanglement with fungal mycelium.

More work could be done on the relative importance of bacterial polysaccharides in soil stabilisation; but as more strains of soil bacteria are examined, a great variety of polymers have been discovered, and it is in the field of structural studies that perhaps the best opportunities for interesting lines of research will be found.

REFERENCES

1. ACTON, C. J., PAUL, E. A. and RENNIE, D. A. Measurement of polysaccharide content of soils, *Canadian Journal of Soil Science*, **43**, 141–150 (1963)
2. ACTON, C. J., RENNIE, D. A. and PAUL, E. A. The relationship of polysaccharides to soil aggregation, *Canadian Journal of Soil Science*, **43**, 201–209 (1963)
3. ADAMS, G. A. and YOUNG, R. Capsular polysaccharides of *Serratia marcescens*, *Canadian Journal of Biochemistry*, **43**, 1499–1512 (1965)
4. ADAMS, G. A. and YOUNG, R. Constitution of an extracellular acidic glucomannan from *Serratia marcescens*, *Canadian Journal of Chemistry*, **43**, 2940–2946 (1965)
5. ALLISON, L. E. Effects of micro-organisms on permeability of soil under prolonged submergence, *Soil Science*, **63**, 439–450 (1947)
6. ASPIRAS, R. B., ALLEN, O. N., CHESTERS, G. and HARRIS, R. F. Chemical and physical stability of microbially stabilised aggregates, *Soil Science Society of America Proceedings*, **35**, 283–286 (1971)

7. ASPIRAS, R. B., ALLEN, O. N., HARRIS, R. F. and CHESTERS, G. The role of micro-organisms in the stabilisation of soil aggregates, *Soil Biology and Biochemistry*, **3**, 347–353 (1971)

8. BAROOAH, P. P. and SEN, A. Nitrogen fixation in *Beijerinckia* in relation to slime formation, *Archiv für Mickrobiologie*, **48**, 381–385 (1964)

9. BJÖRNDAL, H., ERBING, C., LINDBERG, B., FÅHRAEUS, G. and LJUNGGREN, H. Studies on an extracellular polysaccharide from *Rhizobium meliloti*, *Acta Chemica Scandinavica*, **25**, 1281–1286 (1971)

10. BUSH, J. A. and WILSON, P. W. A non-gummy chromogenic strain of *Azotobacter vinelandii*, *Nature*, **184**, 381 (1959)

11. CHESTERS, G., ATTOE, O. J. and ALLEN, O. N. Soil aggregation in relation to various soil constituents, *Soil Science Society of America Proceedings*, **21**, 272–277 (1957)

12. CLAPP, C. E. and DAVIS, R. J. Properties of extracellular polysaccharides from *Rhizobium*, *Soil Biology and Biochemistry*, **2**, 109–117 (1970)

13. CLAPP, C. E., DAVIS, R. J. and WAUGAMAN, S. H. The effect of Rhizobial polysaccharides on aggregate stability, *Soil Science Society of America Proceedings*, **26**, 466–469 (1962)

14. CLAUS, D. 2-Keto-3-deoxygalactonic acid as a constituent of an extracellular polysaccharide of *Azotobacter vinelandii*, *Biochemical and Biophysical Research Communications*, **20**, 745–751 (1965)

15. COHEN, G. H. and JOHNSTONE, D. B. Extracellular polysaccharides of *Azotobacter vinelandii*, *Journal of Bacteriology*, **88**, 329–338 (1964)

16. COHEN, G. H. and JOHNSTONE, D. B. Capsular polysaccharide of *Azotobacter agilis*, *Journal of Bacteriology*, **88**, 1695–1699 (1964)

17. CORPE, W. A. The extracellular polysaccharide of gelatinous strains of *Chromobacterium violaceum*, *Canadian Journal of Microbiology*, **6**, 153–163 (1960)

18. DEDONDER, R. A. and HASSID, W. Z. The enzymatic synthesis of a (β-1,2-)-linked glucan by an extract of *Rhizobium japonicum*, *Biochimica et Biophysica Acta*, **90**, 239–248 (1964)

19. DEINEMA, M. H. and ZEVENHUISEN, L. P. T. M. Formation of cellulose fibrils by Gram-negative bacteria and their role in bacterial flocculation, *Archiv für Mikrobiologie*, **78**, 42–57 (1971)

20. DUGUID, J. P. The demonstration of bacterial capsules and slime, *Journal of Pathology and Bacteriology*, **63**, 673–685 (1951)

21. FINCH, P., HAYES, M. H. B. and STACEY, M. Studies on soil polysaccharides and on their interaction with clay preparations, in *Soil Chemistry and Fertility. Meeting of Commissions II and IV of the International Society of Soil Science* (edited by G. V. Jacks), 19–32. Aberdeen (1966)

22. FORSYTH, W. G. C. and WEBLEY, D. M. Polysaccharides synthesised by aerobic mesophilic, spore-forming bacteria, *Biochemical Journal*, **44**, 455–459 (1949)

23. FREUDENBERG, K. and CRAMER, F. Über die Schardinger-Dextrine aus Stärke, *Berichte der Deutschen Chemischen Gesellschaft*, **83**, 296–304 (1950)

24. FUCHS, A. Synthesis of levan by pseudomonads, *Nature*, **178**, 921 (1956)

25. GORIN, P. A. J. and SPENCER, F. J. T. Isolation of 4,6-O-(1'-carboxyethylidene)-D-galactose from the extracellular polysaccharide of *Corynebacterium insidiosum*, *Canadian Journal of Chemistry*, **42**, 1230–1232 (1964)

26. GORIN, P. A. J. and SPENCER, J. F. T. Exocellular alginic acid from *Azotobacter vinelandii*, *Canadian Journal of Chemistry*, **44**, 993–998 (1966)

27. GORIN, P. A. J., SPENCER, J. F. T. and WESTLAKE, D. W. S. The structure and resistance to methylation of 1,2-β-glucans from species of *Agrobacteria*, *Canadian Journal of Chemistry*, **39**, 1067–1073 (1961)

28. GREENLAND, D. J. Interaction between clays and organic compounds in soils. Part I. Mechanisms of interaction between clays and defined organic compounds, *Soils and Fertilizers*, **28**, 415–425 (1965)

29. GREENLAND, D. J., LINDSTROM, G. R. and QUIRK, J. P. Organic materials which stabilise natural soil aggregates, *Soil Science Society of America Proceedings*, **26**, 366–371 (1962)
30. GRIFFITHS, E. Micro-organisms and soil structure, *Biological Reviews*, **40**, 129–142 (1965)
31. GRIFFITHS, E. and BURNS, R. G. Interaction between phenolic substances and microbical polysaccharides in soil aggregation, *Plant and Soil*, **36**, 599–612 (1972)
32. HARRIS, R. F., ALLEN, O. N., CHESTERS, G. and ATTOE, O. J. Evaluation of microbial activity in soil aggregate stabilisation and degradation by the use of artificial aggregates, *Soil Science Society of America Proceedings*, **27**, 542–545 (1963)
33. HARRIS, R. F., CHESTERS, G. and ALLEN, O. N. Dynamics of soil aggregation, *Advances in Agronomy*, **18**, 107–169 (1966)
34. HARRIS, R. F., CHESTERS, G., ALLEN, O. N. and ATTOE, O. J. Mechanisms involved in soil aggregate stabilisation by fungi and bacteria, *Soil Science Society of America Proceedings*, **28**, 529–532 (1964)
35. HAUG, A. and LARSEN, B. An extracellular polysaccharide from *Berjerinckia indica*, containing L-guluronic acid residues, *Acta Chemica Scandinavica*, **24**, 1855–1856 (1970)
36. HAUG, A. and LARSEN, B. Biosynthesis of alginate. Part II. Polymannuronic acid C-5-epimerase from *Azotobacter vinelandii* (Lipman), *Carbohydrate Research*, **17**, 297–308 (1971)
37. HUMPHREY, B. A. Occurrence of 4-O-methyl glucuronic acid in *Rhizobium* gums, *Nature*, **184**, 1802 (1959)
38. JEANES, A., KNUTSON, C. A., PITTSLEY, J. E. and WATSON, P. R. Extracellular polysaccharide produced from glucose by *Arthrobacter viscosus* NRRL B-1973: Chemical and physical characterisation, *Journal of Applied Polymer Science*, **9**, 627–638 (1965)
39. KOHL, R. A. and TAYLOR, S. A. Hydrogen bonding between the carbonyl group and Wyoming bentonite, *Soil Science*, **91**, 223–232 (1961)
40. LARSEN, B. and HAUG, A. Biosynthesis of alginate. Part I. Composition and structure of alginate produced by *Azotobacter vinelandii*, *Carbohydrate Research*, **17**, 287–296 (1971)
41. LAWSON, G. J. and STACEY, M. Immunopolysaccharides. Part I. Preliminary studies of a polysaccharide from *Azotobacter chroococcum*, containing a uronic acid, *Journal of the Chemical Society*, 1925–1931 (1954)
42. LJUNGGREN, H. and FÅHRAEUS, G. Effects of *Rhizobium* polysaccharide on the formation of polygalacturonase in lucerne and clover, *Nature*, **184**, 1578–1579 (1959)
43. LOPEZ, R. and BECKING· J. H. Polysaccharide production by *Beijerinckia* and *Azotobacter*, *Microbiologia Española*, **21**, 53–75 (1968)
44. MARTIN, J. P. Micro-organisms and soil aggregation. II. Influence of bacterial polysaccharides on soil structure, *Soil Science*, **61**, 157–166 (1946)
45. MARTIN, J. P. Decomposition and binding action of polysaccharides in soil, *Soil Biology and Biochemistry*, **3**, 33–41 (1971)
46. MARTIN, J. P., ERVIN, J. O. and SHEPHERD, R. A. Decomposition and binding action of polysaccharides from *Azotobacter indicum (Beijerinckia)* and other bacteria in soil, *Soil Science Society of America Proceedings*, **29**, 397–400 (1965)
47. MARTIN, J. P., ERVIN, J. O. and SHEPHERD, R. A. Decomposition of iron, aluminium, zinc and copper salts or complexes of some microbial and plant polysaccharides in soil, *Soil Science Society of America Proceedings*, **30**, 196–200 (1966)
48. MARTIN, J. P., MARTIN, W. P., PAGE, J. B., RANEY, W. A. and DE MENT, J. D. Soil aggregation, *Advances in Agronomy*, **7**, 1–37 (1955)

49. MARTIN, J. P. and RICHARDS, S. J. Decomposition and binding action of a polysaccharide from *Chromobacterium violaceum* in soil, *Journal of Bacteriology*, **85**, 1288–1294 (1963)

50. MEHTA, N. C., STREULI, H., MÜLLER, M. and DEUEL, H. Role of polysaccharides in soil aggregation, *Journal of the Science of Food and Agriculture*, **11**, 40–47 (1960)

51. PARIKH, V. M. and JONES, J. K. N. The structure of the extracellular polysaccharide of *Azotobacter indicum*, *Canadian Journal of Chemistry*, **41**, 2826–2835 (1963)

52. PORTOLES, A., LOPEZ, R. and HILDAGO, A. The action of antibiotics on heteroglucan production by strains of *Azotobacteriaceae*, *Canadian Journal of Microbiology*, **14**, 763–767 (1968)

53. POSTGATE, J. R. *The Chemistry and Biochemistry of Nitrogen Fixation* (edited by J. R. Postgate), 174–175. Plenum Press, London (1971)

54. PRATT, Y. T., KONETZKA, W. A., PELCZAR, M. P., Jr. and MARTIN, W. H. Biological degradation of lignin. V. Polysaccharide synthesis from alpha-conidendrin, *Applied Microbiology*, **1**, 171–174 (1953)

55. RENNIE, D. A., TRUOG, E. and ALLEN, O. N. Soil aggregation as influenced by microbial gums, level of fertility and kind of crop, *Soil Science Society of America Proceedings*, **18**, 399–403 (1954)

56. ROUATT, J. W. and KATZNELSON, H. A study of the bacteria on the root surface and in the rhizosphere soil of crop plants, *Journal of Applied Bacteriology*, **24**, 164–171 (1961)

57. RUSSELL, SIR E. JOHN. *Soil Conditions and Plant Growth*, 8th edition, 397. Longmans, Green, London (1950)

58. SALOMON, M. Soil aggregation—Organic matter relationships in redtop–potato rotations, *Soil Science Society of America Proceedings*, **26**, 51–54 (1962)

59. SCHARDINGER, F. Ueber die Bildung Kristallisierter, Fehlingsche Lösung nicht reduzierender Körper (Polysaccharide) aus Stärke durch mikrobielle Tätigkeit, *Zentralblatt für Bakteriologie, Parasitenkunde und Infektionskrankheiten*, **22**, 98–103 (1909)

60. SLONEKER, J. H. and ORENTAS, D. G. Exocellular bacterial polysaccharide from *Xanthomonas campestris* NRRL B-1459 Part II. Linkage of the pyruvic acid, *Canadian Journal of Chemistry*, **40**, 2188–2189 (1962)

61. SMITH, E. J. Purification and properties of an acidic polysaccharide isolated from *Achromobacter georgiopolitanum*, *Journal of Biological Chemistry*, **243**, 5139–5144 (1968)

62. SOUMARE, S. and BLONDEAU, R. Caracteristiques microbiologiques de sols de la region du nord de la France: Importance des 'Arthrobacter', *Annales de l'Institut Pasteur (Paris)*, **123**, 239–249 (1972)

63. SPENCER, J. F. T. and GORIN, P. A. J. The occurrence in the host plant of physiologically active gums produced by *Corynebacterium insidiosum* and *Corynebacterium sepedonicum*, *Canadian Journal of Microbiology*, **7**, 185–188 (1961)

64. STEVENSON, F. J., CLAPP, C. E. and MOLINA, J. A. E. Occurrence of amino sugars in extracellular polysaccharides of the genus *Rhizobium* and some observations regarding their distribution in somatic components, *Soil Science Society of America Proceedings*, **34**, 759–764 (1970)

65. SUTHERLAND, I. W. Phage-induced fucosidases hydrolysing the exopolysaccharide of *Klebsiella aerogenes* Type 54[A3(S1)], *Biochemical Journal*, **104**, 278–285 (1967)

66. SUTHERLAND, I. W. Bacterial exopolysaccharides, *Advances in Microbial Physiology*, **8**, 143–213 (1972)

67. SUTHERLAND, I. W. and WILKINSON, J. F. The exopolysaccharide of *Klebsiella aerogenes* A3(S1)(Type 54). The isolation of *O*-acetylated octasaccharide, tetrasaccharide and trisaccharide, *Biochemical Journal*, **110**, 749–754 (1968)

109

68. SWABY, R. J. The relationship between micro-organisms and soil aggregation, *Journal of General Microbiology,* **3,** 236–254 (1949)
69. WEBLEY, D. M., DUFF, R. B., BACON, J. S. D. and FARMER, U. C. A study of polysaccharide-producing organisms occurring in the root region of certain pasture grasses, *Journal of Soil Science,* **16,** 149–157 (1965)
70. WEGEMER, D. E. and GAINER, C. Studies on a psychrophilic bacterium causing ropiness in milk. II. Chemical nature of capsular polysaccharide, *Applied Microbiology,* **2,** 97–99 (1954)
71. WHISTLER, R. L. and KIRBY, K. W. Composition and behaviour of soil polysaccharides, *Journal of the American Chemical Society,* **78,** 1755–1759 (1956)
72. WILKINSON, J. F. The extracellular polysaccharides of bacteria, *Bacteriological Reviews,* **22,** 46–73 (1958)
73. YOUNG, R. and ADAMS, G. A. Structural features of two extracellular polysaccharides from *Serratia marcescens, Canadian Journal of Chemistry,* **43,** 2929–2939 (1965)
74. ZEVENHUISEN, L. P. T. M. Methylation analysis of acidic exopolysaccharides of *Rhizobium* and *Agrobacterium, Carbohydrate Research,* **26,** 409–419 (1973)

6 *Rhizobium* in the Soil

P. S. NUTMAN

6.1 Introduction

The chief interest and importance of *Rhizobium* lies in its nitrogen-fixing activity in the root nodule, but it is also a normal soil inhabitant and this chapter will be restricted to this part of its life cycle. An attempt will be made to bring together many aspects usually considered separately; to provide background it will be necessary to review briefly the taxonomy and general properties of *Rhizobium* in culture. Finally, we shall consider the microbiological problems associated with the introduction ('inoculation') of *Rhizobium* into agricultural soils for improvement of pasture and grain legumes.

6.2 The classification and genetics of *Rhizobium*

Even in the limited context of the soil, the literature on *Rhizobium* is probably more extensive than that of any other soil organism, almost rivalling that on *Escherichia coli,* with which it has some affinity, as an object of study. For this reason only the more important or recent references will be quoted. Like *Escherichia,* it has suffered several changes in name since its original isolation in 1888 as *Bacillus radicicola,* and the definitions of generic and specific taxa are still in dispute. Some microbiologists would retain the overriding importance of symbiotic affinities, whereas others do not place these characteristics above many others that can be used.

The current host-based classification of a single genus and six named species, viz. *Rhizobium leguminosarum* Frank, *R. trifolii* Dang., *R. meliloti* Dang., *R. lupini* Schrod., *R. phaseoli* Dang. and *R. japonicum* Kirch., is unsatisfactory because the cross-inoculation groups of host plants are very disparate in size (the clover group consists of *Trifolium* species only, whereas the cow-pea miscellany comprises several hundreds of genera) and also because cross-infection is often poorly defined and sometimes anomalous.

Much comparative biochemistry, serology, antibiotic and phage sensitivity of strains of *Rhizobium* isolated from different hosts suggests the consolidation of *R. trifolii* with *R. leguminosarum* and *R. phaseoli* (as *R. leguminosarum*) and *R. lupini* with *R. japonicum* (as *R. japonicum* or, less favoured, *Polymyxa*). *R. meliloti* remains a distinct species. Each of these three major groups can be further subdivided; the cow-pea group to a very considerable extent on the basis of patterns of virulence.[25, 26, 32] Some of the above studies also dealt with strains of *Agrobacterium, Arthrobacterium, Azotobacter, Bacillus, Beijerinckia, Chromobacterium, Flavobacterium, Serratia* and *Vibrio. Agrobacterium* appears to be closely related to *Rhizobium*.

Genetic studies have also thrown light on *Rhizobium* taxonomy and on the mechanisms of gene transfer. These generally confirm Graham's groupings, but Gibbins and Gregory[29] show that *R. trifolii* and *R. leguminosarum* can be distinguished by DNA hybridisation, whereas *R. meliloti* and *R. phaseoli* show some relationship. DNA base ratio analysis and DNA hybridisation also show that *Agrobacterium tumifaciens* and *R. leguminosarum* are closely related. The nodule bacteria of the tropical pasture plant *Lotononis bainesii* differ markedly from other rhizobia; they also have an abnormally high % G + C (% guanine + cytosine).[31] The nodule bacteria of the non-legume *Trema cannabina* appear to be closely related to *R. japonicum.*[81] *Serratia marcescens* DNA has a low hybridisation rate with *Rhizobium* DNA; its % G + C is similar to that of *R. trifolii* DNA. Transformation is claimed between *Rhizobium* and *Azotobacter* and between *Rhizobium* and *Agrobacterium,* and some affinity is reported between rhizobia and the bacteria that form leaf nodules on the *Rubiaceae.* That *R. trifolii* and *R. leguminosarum* are closely related is underlined by reports of sporadic infection of infection of clover by pea nodule bacteria and by the recent demonstration that the incidence of red clover by pea strains can be greatly increased by plant breeding.[36] The low frequency of infection of peas by *R. trifolii* can be increased by mutation using UV, X-rays, fast neutrons or antibiotics.[69]

Legume[20] and non-legume[78] tissue culture protoplasts can take up *Rhizobium* cells, but the claim of transfer of labelled proteins from rhizobia to the host's organelles in infected cells is unconfirmed. These observations do not necessarily indicate genetic homology; indeed the very low combining rate (less than 0.6%) of DNA from

112

soybeans and *R. japonicum* (Rake[61]) is such that any transfer of genetic material would be quite unstable. Variants that occur naturally or can be induced by untoward physical or chemical conditions, irradiation, chemical mutagens or metabolic inhibitors have been used widely as markers in studying recombination, transformation, plasmid inheritance and transduction and in ecological work. Phage-induced changes among *Rhizobium* strains, due to either selection or lysogeny, seem to be more the rule than the exception; phages from soil are very variable in morphology and host range.[4, 44] The possible conjugation of *Rhizobium* in soil has not been examined, although Weinburg and Stocky[86] have demonstrated conjugation and recombination in *Escherichia coli* in soil. Kowalski[46] has suggested that transduction by phage may be important in soil. The parts played by these mechanisms of genetic change in modifying natural populations and their evolution is wholly speculative.[24]

The characteristics of Rhizobium

Rhizobium is a small to medium-sized $(0.5–0.9 \times 1.2–3.0\,\mu m)$ non-spore-forming, Gram-negative rod, usually motile by peritrichous flagella or a single flagellum. In soil it occurs as rods or small cocci, with or without flagellae, the proportion of forms altering with conditions, and these were considered at one time to constitute a life cycle. As an aerobic heterotroph it is easily grown on a range of bacteriological media that contain yeast extract as a source of accessory growth factors, although wholly synthetic media have been devised (see, for example, reference 72). Actidione may be added to media to discourage fungal contamination, and Congo Red, which is only slowly absorbed, is useful in media for plating. *Rhizobium* requires trace amounts of Ca, Mg, Zn, Co and possibly Ni, but is inhibited by heavy metals at above trace concentrations, Ni but not Zn interfering with Co uptake or metabolism. Rhizobia are remarkably tolerant to high concentrations of Cr and Mn.

Rhizobium survives at low temperature and can be freeze dried but is soon killed at temperatures above about 50°C for more than a few hours[23] or by drying. Much is known about its sensitivity to antibacterial substances, herbicides, fungicides, pesticides, soil fumigants and other agricultural chemicals (see, for example, references 28, 42, 43 and 83) but less is known about the degradation of such compounds by *Rhizobium*.[52]

Growth is usually best at neutral or slightly acid pH. Most strains produce acid in culture but this is not related to the acidity of the soil of origin.[40] *Rhizobium* produces plant growth substances such as indole derivatives, cytokinins and traces of gibberellic acid, although more is produced in the nodule than in culture, and some strains excrete B vitamins, plant-wilting factors, bacteriocins and porphyrins.

Rhizobium produces extracellular polysaccharide gums which differ somewhat between species and strains in composition and serology (see, for example, reference 91). Cellulose fibrils but not cellulase are also produced by *Rhizobium*.

The movement of *Rhizobium* in soil is governed by moisture and ceases when the water film lining the soil pores becomes discontinuous,[34] which occurs normally at water tensions still adequate for seed germination. At good moisture levels records of movement in sand and in soils range from 2 to 20 mm per day.

6.3 Counting *Rhizobium* in soil

Because *Rhizobium* is similar culturally and biochemically to many other soil organisms, it cannot easily be distinguished on soil plates and much work has therefore gone into devising selective media using dyes or mixtures of antibacterial substances.[73] These are of very limited use and, for counting, recourse is had to serological methods, at first using agglutination or precipitation reactions and later gel diffusion,[25] or labelling with fluorescent antibody.[41] For identifying nodule isolates the fluorescence technique is unrivalled, but for soil counts non-specific absorption and, to a lesser extent, autofluorescence, reduce the accuracy of the method. *Rhizobium* mutants selected for resistance to streptomycin or kanamycin can be used in recovery plate counts and have proved satisfactory for studies on competition, etc.[57] A mutant strain of *R. japonicum* that induces chlorosis in the host has been used to study strain competition.

More reliable counts can be made by using the host plant in a biological assay in which dilutions made from a weighed quantity of soil are added to sterile-grown plants. Any rhizobia present multiply rapidly in the root surroundings (rhizosphere), so that all dilutions containing at least one organism will thus nodulate the test plant. This method was first used by Wilson,[88] who calculated the number present by simple proportion, but today a probability factor is incorporated into the calculation (the 'most probable number' estimate). Small-seeded host species such as clovers or medics can be grown in test tubes on an agar medium or in sand–vermiculite mixtures, and larger hosts in Leonard jars or plastics pouches. Recovery counts show these methods to be generally good for *R. trifolii* and *R. lupini* but less so for *R. leguminosarum* and *R. meliloti,* possibly through microbial predation or antibiosis[56] affecting some *Rhizobium* spp. more than others.

6.4 *Rhizobium* ecology

The rhizosphere

Rhizobium is pre-eminently a rhizosphere organism multiplying on the root surfaces and in the root surroundings, and within the mucigel layer of both legumes and non-legumes, but especially legumes.[67] The different species and strains are stimulated to slightly different degrees but not always more by the hosts they infect than by non-susceptible hosts. Stimulation is greatest at places where lateral roots emerge, and generally extends 10–20 mm from the root surface into the soil. It varies seasonally but is not much influenced by temperature, light, foliar sprays or fertilisers. Many other organisms share in this stimulation in the legume rhizosphere but Gram-negative aerobic bacteria increase most. Fungi and, at first, cellulose decomposers also increase, but anaerobes and actinomycetes are not much affected.

Increased growth of *Rhizobium* in the rhizosphere is a response to excretion of nutrients. Plant roots exude, or lose through plasmoptysis, a great variety of substances: energy sources, amino acids, growth factors, especially B group vitamins, and enzymes. That the rhizosphere stimulation is a response to a complex mixture of substances was demonstrated by Rovira,[66] who was unable to replace fully the stimulating effect of root exudates by mixtures of glucose, soil extract, amino acids and all growth factors known to be excreted by roots.

Effect of soil acidity, alkalinity and salt content

The species and strains of *Rhizobium* are differently affected by acidic or alkaline soil conditions. Most strains of *R. trifolii* do not grow below about pH 5.0 or above pH 7·0 but *R. meliloti* strains can tolerate for limited periods pHs as low as 3.5 and as high as 8.7. *R. meliloti* and *R. leguminosarum* seem less sensitive to alkaline conditions than *R. japonicum*. *R. trifolii* can tolerate pH as low as 3.5 and its properties remain unchanged even when held at this pH over long periods.[84] Wilson[89] studied *R. trifolii* and *R. leguminosarum* in three soils under arable cultivation which had received a range of chemical treatments involving the addition of sulphur, sulphates or carbonates over a 40 year period, which changed the salt concentration and the pH over the range 4.8–8.0. Fewest *R. trifolii* and *R. leguminosarum* were found in those soils rendered most acid. *Rhizobium* was shown to be intolerant of salinity, especially that caused by more than 0.4% $NaHCO_3$; K, Na, SO_4 and Cl ions seem less harmful and NaCl is less harmful than Na_2SO_4n, K_2SO_4 or KCl. The growth of *R. japonicum* is strongly retarded in media containing 0.008 M NaCl and does not grow with 0.16M NaCl;[90] nitrogenous fertilisers are less injurious to *Rhizobium* than NaCl. Bhardwaj[8] found that strains from *Sesbania cannabina* growing in a saline–alkaline soil are more tolerant of these conditions than strains from a normal soil. Subba Rao *et al.*[79] found *Rhizobium meliloti* more tolerant of salt than its host plant.

Moisture and temperature

In sterile soil cultures allowed to dry slowly, rhizobia can survive for decades. *R. meliloti* was re-isolated from such cultures originally made up with 0.5% mannitol after 45 years,[38] and *R. japonicum* without addition of sugar up to 19 years.[71] When dried on to glass beads, death occurs within days or weeks, 60% relative humidity being more damaging than either 0 or 20% R.H. *Rhizobium* from broth culture dried on seeds or in peat also survives only for a short time and this is accentuated by exposure to sunlight. Strains differ in survival, *R. trifolii* surviving better than *R. meliloti* or *R. lupini*. Among a range of non-spore-forming soil organisms, Chen and Alexander[16] found *Rhizobium* most susceptible to desiccation, requiring for any growth a minimum water activity, a_w, of 0.990–0.993. Survival on exposure to desiccation and high temperature is better in heavy or organic soils than in light soils, and is improved in soils amended with montmorillonite, illite, etc., the clay probably interacting with the bacteria to reduce the rate of water loss.[48] The protective effect of clays may be related to the nature of the contact with the bacteria and this varies from strain to strain. By electrophoretic studies Marshall[49] showed that there is an edge-to-face rather than a face-to-face association. Slow-growing rhizobia tend to have surface acidic groups (carboxyl) exposed, whereas fast growers have more complex structures, including exposed amino groups. The humic and fulvic acid content of soils may also have an ameliorating effect, because both appreciably improve growth when added to culture media, although ^{14}C-labelling of humic acid showed that this substance was not taken up by *R. trifolii*.[9] In moist soil, strains are less tolerant of high temperature than in drier soils, although damagingly high temperatures are most often met when soils are dry. Strains from different climatic regions may differ by as much as 3 or 4°C in their optimum growth temperatures. Wilkins[87] reports greater temperature tolerance in strains isolated from naturally occurring legumes of the hot western plains of New South Wales, Australia, than in strains from the cooler tablelands, but there is no evidence of adaptation among strains introduced into these regions.

Chatel and Parker[15] demonstrated the poorer persistence of *R. trifolii* introduced into the light soils of the dry regions of Western Australia compared with naturally occurring *R. lupini* strains. In this area temperatures of 60°C were commonly recorded at 1.3 cm depth in the soil. These soils lack the protective action of a sufficiently large clay component. By subjecting cultures to 70° for short periods Delin[22] isolated strains that died more slowly at this temperature than the parental strain, but the limited variations in temperature-resistance among rhizobia from contrasting climates do not offer much hope of improved temperature tolerances. Claims to have isolated rhizobia that are very resistant to heat (100°), UV light and radiation or are pigmented remain unconfirmed.

Other soil factors affecting Rhizobium

Reference is made above to the differing susceptibilities of species and strains to antibacterial agents, most of which are of soil microbial origin, and there is much circumstantial evidence that *Rhizobium* in soil is influenced by microbial antagonism. Actinomycetes are more antagonistic to rhizobia than bacteria and some inhibit nodulation. Hattingh and Louw[35] examined the responses of several strains of *R. trifolii* individually to more than 1000 isolates of fungi and bacteria from the rhizosphere and rhizoplane of clovers growing in South Africa; 8% were antagonistic and 16% were stimulatory; pseudomonads were prominent among the antagonists. In parts of Western Australia where *R. trifolii* is difficult to establish in soil about 14% of isolates of fungi and bacteria were antagonistic to this species; water extracts of these soils were also toxic. Among aerobic spore-formers from South American soils that inhibited rhizobia was a strain of *Bacillus subtilis,* which produced an antagonistic peptide (rhizobacidin) that inhibited the growth of all Gram-positive organisms tested but not any Gram-negatives except *Rhizobium.* Anderson[3] compared the effects of a range of soil fungi and bacteria on *Rhizobium* growing in pure culture and symbiosis. Some were beneficial and others harmful in both situations. One fungus inhibited *Rhizobium* only in the host rhizosphere. Krassilnikov and Koreniako[47] also examined the influence of *Pseudomonas* and *Achromobacter* species on the growth of *R. trifolii* in the rhizosphere and found some inhibitory, some stimulatory and others with no effect. Of the many species of fungi associated with nodules, some inhibit growth of *Rhizobium* in the rhizosphere.[17]

Schwinghamer[70] studied antagonism among 300 strains of rhizobia using six test *Rhizobium* strains; 35% showed mild antagonism (due to a dialysable substance) towards two of the test strains and 8% were lysogenic by producing a bacteriocin-like substance.

Rhizobium bacteriophages occur widely in soils[5, 45] and at one time were thought to cause soil sickness of lucerne and other legumes by reducing numbers below that optimal for infection. However, the occurrence of rhizobial phages is as widespread as their host bacteria and perhaps related to *Rhizobium* populations; without special aseptic precautions pot experiments with nodulated legumes quickly become contaminated by phage.[45]

Phage-resistant forms arise by mutation and may differ in other characteristics. Because transduction is likely to occur in soil, genetic effects of phage may be more important than effects on number.

Rhizobium can be parasitised by *Bdellovibrio* but how widespread this may be is not yet known. Sullivan and Casida[80] did not find *Bdellovibrio* that attack rhizobia in filtrates from a cornfield or from a sewage effluent pond in Pennsylvania, USA, but Parker and Grove[58] found *Bdellovibrio* in Western Australian soils. Their filtrates were active against *R. meliloti, R. trifolii, Agrobacterium tumifaciens* and *Agrobacterium radiobacter; R. lupini,* however, was not lysed. *Rhizobium* strains vary in their susceptibility to lysis by soil myxobacteria. About half of the strains studied by Singh[76] were partly or completely lysed. Most strains of *Rhizobium* are partly or completely eaten by the giant amoeboid organism (*Leptomyxa reticulata*), but of 16 strains examined only 2 were partly eaten by large soil amoebae, none were partly eaten by small amoebae and 3 were partly eaten by flagellates.[75, 77] Variation in edibility between strains was not correlated with the amount of slime produced by them.

Numbers and distribution in soil

Rather less than 10% of more than 12 000 known species of leguminous plants have been examined for root nodules; and of these, relatively few (mostly among the *Caesalpinioideae*) are not nodulated and remain so when inoculated with *Rhizobium* strains taken from nodulating members of the same plant families.[1] This indicates not only that resistance in these legumes is probably a host property but also that distribution of rhizobia, in general, corresponds with host distribution; where a susceptible legume is present, so are its nodule bacteria. The converse is more difficult to establish but it seems that, in general, rhizobia are absent where their hosts are not found. The few exceptions are of very small populations of rhizobia which may have come from elsewhere—in dust, on implements, etc.

Australian workers found *Rhizobium meliloti* well distributed in natural pastures in the McQuarie region of New South Wales, which in the past contained no species of *Medicago* or *Melilotus,* and in 25 out of 26 sites in New South Wales and Queensland, where medicks now occur; numbers varied from 10 to 10^6 organisms per gram of dry soil. There appears to be no relationship between the distribution of rhizobia and topographical features of the landscape, except in so far as they affect the distribution of the host species. Similar results were found for the distribution of *R. trifolii* in naturally seeded hill pastures of the western uplands of the U.K. Except for two strains isolated from sites between 800 ft and 900 ft, Masterson[50] found no rhizobia or clovers above 700 ft in south-east Ireland. There, and in Welsh and Scottish hills, the strains isolated from the higher altitudes tended to be poorly effective.[37] At some of these sites poor effectiveness was associated with low pH and low base status in the soil.

Nutman and Ross[56] surveyed the populations of rhizobia in certain of the differently manured plots of the Park Grass Experiment at Rothamsted, which for more than a century has been meadowland cut for hay. There was a striking correspondence between the occurrence of *Rhizobium* species in the soil of each plot and the presence of its host legume. *Medicago* and *Melilotus* were absent from the entire field, and in no plot was *Rhizobium meliloti* found, although in neighbouring fields *Medicago lupulina* is a common weed. *Rhizobium trifolii, R. leguminosarum* and *R. lupini* were present in most plots, *R. trifolii* being most abundant—especially in limed plots, where numbers commonly exceeded 10^6 per gram. *Rhizobium trifolii* also showed a rhizosphere effect even after so long a period of clover cover, especially in the somewhat acid plots. Rhizobia were most abundant in the 0–1 in sample, but in most plots appreciable populations were found to the maximum depth sampled (12 in). Rhizobia were absent in plots with pH values less than about 4, and surviving *R. trifolii* in the soil of marginally low pH tended to be poorly effective in nitrogen fixation on *Trifolium pratense,* but neither acidity nor absence of lime was invariably associated with decline in effectiveness.

Nitrogen fertilisers affect the soil populations of rhizobia mainly through their influence on the distribution of host species. They do not much affect the symbiotic characteristics of the strains: all strains isolated from limed plots given nitrogenous fertilisers for more than 100 years were of good effectiveness.

De Escuder[21] counted rhizobia in a permanent pasture contain-in white clover, sown prior to 1956 in Kent on a gley soil at pH 6.7, and found more than 10^6 *R. trifolii* per gram, 10^4–10^5 *R. leguminosarum* per gram, about 10^2 *R. lupini* per gram and no *R. meliloti.*

Nodule bacteria in arable soils (including sown pasture species) have been more studied. The first extensive studies were made by Wilson,[88, 89] who followed their development in the first legume crop (clover, peas and cowpea) following wheat in Midwestern USA. Counts were very variable, ranging from less than 1 organism per gram to more than 10^5 per gram. Clovers stimulated the multiplication of *R. trifolii* more than *R. leguminosarum.*

Cow-peas were least stimulatory for *R. trifolii,* which, however, eventually became more abundant than *R. leguminosarum,* irrespective of cropping. Numbers tended to be minimal in winter and to increase parallel with the growth of the host. Heavy farmyard manuring of a silty loam increased numbers even under fairly acid conditions (ph 5.2–5.8), and lime was stimulatory. Few rhizobia were found under oats or potatoes.

Peterson and Gooding[60] determined the presence or absence of *R. meliloti* in more than 300 ten-gram samples of Nebraskan soils. *Rhizobium meliloti* was only found under non-acid conditions (related to soil type) and was not influenced by carbonate, phosphate, total base exchange capacity or exchangeable calcium.

Jensen[39] sampled agricultural soils at 214 sites in Denmark and found *R. trifolii* almost ubiquitous, even in two very acid soils (pH 4.3 and 4.4) and in some heath and forest soils. An average of 10^5 organisms per gram of soil was counted in 27% of soils not carrying clover. *R. meliloti* was present in 70% of soils with pH above 6.0 (generally more than 10^5 per gram). When inoculated into these soils under laboratory conditions, *R. meliloti* died rapidly at pH below 5.0, died more slowly at pH 5.2–5.7 and survived for 22 weeks above pH 5.9.

Nutman[54] and Nutman and Ross[56] counted the four most common *Rhizobium* species in several classical arable fields of the Rothamsted (clay loam) and Woburn (sandy loam) soils. Numbers of each species were few (or sometimes absent) in soils continuously cropped to non-legumes or in rotations without legumes or in long-term fallows. In Barnfield (Rothamsted, continuous roots since 1843), with no records of leguminous weeds, *R. trifolii* and *R. leguminosarum* averaged a few hundreds of cells per gram of soil, and *R. meliloti* and *R. lupini* were scarcer, or even absent in some samples: these low numbers were unaffected by mineral or organic manuring. In Broadbalk (Rothamsted, continuous wheat since 1843) *R. leguminosarum* was most abundant (about 10^5 per gram); *R. trifolii* and *R. lupini* were sparse (about 10^2 per gram) and *R. meliloti* very sparse (about 10 per gram). Abundance in this field was not related to the occurrence of leguminous weeds. In other Rothamsted fields the numbers of rhizobia were much greater—for example, in a limed field the average populations under *Medicago lupulina, Vicia faba* and *Trifolium pratense* were, respectively, 5×10^5 *R. meliloti,* more than 2×10^6 *R. leguminosarum* and 7×10^5 *R. trifolii* per gram of dry soil.

Similar studies were made at Wye, Kent, by de Escuder.[21] *Rhizobium trifolii* and *R. leguminosarum* were more abundant than *R. lupini* and *R. meliloti,* and, as in other studies, large populations were only found where the respective host had been grown recently. Fertilisers had little effect on numbers. The serology of *Rhizobium* strains was also studied at Wye at sites known to have grown lucerne at different times since 1956 and to have been inoculated with a commercial inoculant prepared from strain 2001 of the Rothamsted collection. None of the 110 isolates was serologically identical with strain 2001, although about half reacted with some of its antigens in gel-precipitation tests. Tuzimura and Watanabe[82] recorded more than 10^4 *R. meliloti* per gram under lucerne and more than 10^5 *R. trifolii* per

121

gram under *Trifolium lupinaster* in Japanese soils. Bell and Nutman[6] counted *R. meliloti* at five sites at or near Rothamsted not known to have previously grown a medick. At three sites (all at pH 6.0) no *R. meliloti* was found. In a calcareous soil (pH 7.5) about 10^2 per gram were counted and at a rather acid site (pH 5.5) the count exceeded 10^3 per gram.

Burton, Allen and Berger[13] found that *R. phaseoli* was not abundant in soils of the Midwestern USA and that numbers were not related to soil reaction.

Weaver, Frederick and Dumenil[85] found a wide range in counts of *R. japonicum* (from less than 10 to more than 10^6 per gram of soil) in soils at 52 sites in Iowa; numbers were related principally to crop history, soil texture and organic matter.

Serological typing is much used to study the distribution of *R. japonicum* strains and the competition between strains already in soil and those introduced by inoculation. Many serological types were identified even within a single locality, and the spectrum of serological types among nodule isolates differed between localities and was affected by soil type, planting date, etc.[7, 14] The influence of the host on rhizobia distribution in South African soils was demonstrated by Scheffler and Louw,[68] who also identified many more antigens among isolates from naturally occurring legumes than among isolates from alien legumes. Ham, Frederick and Anderson[33] observed an effect of pH on serotype, but, in general, strain serology does not seem to be related to soil properties.

Early laboratory and pot experiments on competition between pairs of *Rhizobium* strains of the clover, pea and soybean groups used ones that could be identified culturally or serologically. They demonstrated differences in rates of growth in the rhizosphere which sometimes accounted for the different proportions of strains isolated from the nodule. Using serology, Read[62] and Skrdleta[74] found the proportion of strains in nodules to be simply related to the proportion of cells of each strain in the inoculum; but when inoculation by a second strain was delayed, the proportion of nodules formed by it declined. Other workers, using different strains and conditions, were unable to demonstrate any proportionality between populations in the rhizosphere and the nodules. Robinson[64] and Jones and Russell[41] showed large disparities in these proportions: in one case a difference of 10 000 : 1 was needed to overcome the selective effect of the host in determining which bacteria formed the nodules. Competition between small indigenous populations of effective strains of *R. meliloti* and large inocula of an ineffective strain was examined by Bell and Nutman[6] at five sites on three soils (heavy loam, sandy loam and calcareous soil). At each site the inoculated strain multiplied in the lucerne rhizosphere and initially formed all the nodules, but this strain was more or less completely suppressed in the rhizosphere by effective strains after periods of 12–18 months. Nutman and Read[55] described another effect of the host on the composition of the soil population of strains by showing that local strains of *R. trifolii* (in Sweden and the UK) were slightly more effective with local cultivars of red clover than with cultivars from other areas.

Different patterns of competition in the field have been described by workers using cultural, antibiotic-resistant and serological markers[57, 62] and using strains of *R. japonicum* that induce chlorosis in the host.[51] These also show that the host may have some determinative influence; the nature of this selective effect is obscure and difficult to study because of the complexity of infection and nodule formation.[53]

Without stimulation by the plant root, the numbers of rhizobia in soil decline at rates that depend upon the strain's intrinsic capacity to survive and on the soil's physical, chemical and biological characteristics. Wilson[89] observed that 3 years' fallow followed by 3 years' Timothy grass on a heavily manured silty loam soil in the USA had no effect on the numbers of *R. trifolii* and *R. leguminosarum*. Giltner and Longworthy[30] studied the persistence of non-spore-formers, including *Rhizobium* (then called *Pseudomonas radicicola*) in soil allowed to dry, and found that more survived in heavy than in light soils or where milk, albumin, gum arabic, agar or soluble starch was added. Bonnier[11] compared the survival of three strains of *Rhizobium* in eight different soils. Numbers declined very rapidly in sandy or loess soils, or in soil under *Sequoia;* survival was best in alder wood soil. Nutman and Ross[56] followed the changes in *R. trifolii, R. leguminosarum* and *R. meliloti* in long-term bare fallows at Rothamsted (started in 1960 from permanent grass) and Woburn (started in 1959 after arable). Initially, the numbers of *R. trifolii* and *R. leguminosarum* were similar in the two fields (about 10^5 and 10^3, respectively) but there were more *R. meliloti* at Woburn (more than 10^3) than at Rothamsted (fewer than 10^2). Numbers of *R. trifolii* and *R. leguminosarum* declined similarly at each site but with major fluctuations, especially on the heavier soil. This seemed to be related to summer rainfall and possibly the stimulatory effect of temporary growth of weeds. The decline in numbers of *R. meliloti* was much more rapid than that of *R. trifolii* and *R. leguminosarum*. Zero counts of *R. meliloti* were first recorded after 4 years' fallow at both sites.

6.5 The artificial introduction of rhizobia into soil

The Woburn bare-fallow experiment was later used for growing inoculated lucerne and illustrated the rapid build-up of rhizobia in the rhizosphere corresponding to early seedling development.[6] At sowing time (May) there was less than 1 *R. meliloti* per gram soil, but already in June numbers had increased to more than 10^5 per gram of soil. Numbers increased slowly in the limed plots sown to rye grass, but neither liming nor nitrogenous fertilisers affected numbers in the lucerne rhizosphere. This remarkable dependence of the soil population of *Rhizobium* on host stimulation (augmented under agricultural conditions by the spread and growth of the host, sometimes in monoculture) is sufficient to account for their distribution in natural and agricultural habitats, and for the need to inoculate some legumes in soil where they or closely related species have not been grown. Without this phenomenal multiplication, inoculation on a field scale would need to be so massive as to be impracticable. An ordinary good soil contains up to about 10^{20} bacteria per hectare to a depth of 10 cm, whereas seed inoculation at recommended rates introduces at the most 10^9 *R. trifolii* per hectare on inoculated white clover seed or 10^8 *R. japonicum* per hectare when soybean is inoculated.

At first, soil or broken-up nodulated roots of the legume to be sown was used for inoculation. This was superseded by pure or mixed cultures, grown in liquid or agar media, and generally using a carrier, such as humus and soil, coir dust, soybean meal, rotted maize cobs and vegetable compost, lignite, bagasse, cellulose powder or diatomaceous earth and compost. Peat is now the preferred material, which, because it often contains antagonistic bacteria and actinomycetes, is best sterilised and neutralised before seeding with a bacterial culture and leaving to mature. The inoculated peat culture is applied to seed or soil, sometimes with additives such as bentonite, gums or hemiculluloses. The techniques and problems of inoculation, production and use are reviewed by Date,[19] Roughley[65] and others.

In addition to the many soil factors that affect survival or prevent rapid multiplication of the rhizobia, the inoculum applied to the seed is also susceptible to damage from fertilisers, plant-protective chemicals in seed dressings, toxic seed diffusates and inhibitory substances from the legume seedlings root or even nodule, or originating in nearby plants. Survival of rhizobia on legume seeds may be less than on glass beads but may be increased by pre-soaking seeds.

Certain leguminous seed diffusates are inhibitory to many Gram-positive and Gram-negative organisms; that from *Trifolium repens* has been identified as a mixture of tannins and the flavonol myricetin.[27] Bonnier[10] found that the roots of *Galega officianalis* and *Schotia* species inhibited a strain of *R. meliloti* but not strains of *R. trifolii* and *R. japonicum,* and noted that diffusates from *Medicago sativa, Glycine max, Vicia villosa, Vicia narbonensis* and *Lathyrus luteus* were not inhibitory. Extracts of *Ambrosia elatior, Euphorbia corollata* and *Helianthus annus* (containing polyphenols and gallotannins) and leachates from soil of abandoned fields where these weeds occur inhibited about half of test *Rhizobium* strains and reduced the nodulation of legumes to which they were added.[63] Some of these substances are thought to protect seedlings from root pathogens. These harmful effects can sometimes be avoided by pelleting the inoculated seeds with mixtures of lime, phosphate, clay, etc., or by placing the inoculum below or alongside the seed, or by using heavier inocula. Heavy inoculation (or sometimes successive inoculation) and inoculation of mixtures of *Rhizobium* with other organisms such as lactobacilli or pseudomonads have also been used to improve inoculum establishment and to overcome competition from poorly effective strains or problems of soil acidity.

To avoid rapid death of the inoculum on the seed, attempts have been made to introduce *Rhizobium* into the seed either by stigma inoculation of the host plant or by placing the seed in a broth culture which is evacuated and then brought back to normal pressure, thus forcing culture into seed crevices. However, ^{32}P labelling of the cultures used for vacuum treatments showed that it is almost exclusively the damaged or dead seed that are impregnated with bacteria.[18] Pre-inoculation procedures are not to be recommended, because treated seed often suffers subsequently from fungal spoilage; even pelleting is not always beneficial. In none of these methods is it possible to inoculate the seed heavily with bacteria.

6.6 Completing the circle

Some legumes such as groundnut and subterranean clover bury their seed pods below ground, which thereby become contaminated with soil and rhizobia. The frequent need to inoculate these crops, however, suggests that this method of carryover or dissemination of bacteria, though possibly significant in natural plant communities, is unimportant in agriculture. The pods of herbaceous legumes also often become contaminated with soil, but Brockwell[12] found no nodule bacteria on 84% of a sample of pods of *Medicago tribuloides* and only few on the remainder.

The soil is undoubtedly the major reservoir of free-living rhizobia and under natural conditions seldom needs replenishing. Under dense or pure stands of legumes multiplication in the rhizosphere and release of rhizobia from the nodules liberally recolonises the soil for the succeeding crop.

It was once thought that the nodules' bacteroids broke up to form small cocci which then escaped into the soil, but the great disparity between total and viable counts of nodule suspensions questioned this assumption. Almon,[2] using micromanipulation, transferred more than 400 bacteroid cells singly into one or other of eight different bacteriological media. The only one that grew was discounted as a contaminant. Many studies have since shown that old bacteroids lyse, the decaying nodule becoming colonised from the old infection threads and by organisms from outside. This is a seasonally recurring event in most nodules, but nodules of clovers growing under certain conditions have been shown to survive the winter to resume growth in the spring, and truly perennial nodules of gorse and broom, some showing as many as six annual growth rings, have been described.[59]

With few exceptions, microbes, like snakes and spiders, rank low in popular esteem, and this chapter by omitting all but the barest reference to *Rhizobium*'s most important attribute of N-fixation, has not made a case for including it among the elect, although in carbon cycling, in producing soil-stabilising gums and plant growth factors and in solubilising phosphate it may perform minor useful functions.

Over the last 60 years a significant part of the large increase in the world's primary production has depended upon the large changes brought about in the microbial population of soils by the spread of legume crops and by the practice of inoculation. Most spectacular increases in production have been achieved where competition from naturally occurring rhizobia was lacking and where conditions favoured the process of inoculation, as in the soybean belt in America and in East Europe, or in pastures containing subterranean clover in Australia; the greatest success has normally been with alien plants and alien bacteria. Once inoculation has been successful, the natural processes of rhizosphere colonisation tend to take over, and where this occurs inoculation may be a once-for-all process or only occasionally necessary.

Over the next few decades this 'colonising' phase will continue, especially for the rhizobia of pasture legumes in temperate and warm-temperate regions not yet using sown pastures. Over this period much effort will also probably go into developing satisfactory legume–rhizobia associations for the sub-tropical grasslands and in matching the inoculum more closely to the host's needs. Other developments may be the modification of the soils' microbial population by inoculation or other means under conditions of climatic or other stress, and against competition from ineffective or less effective strains of rhizobia. Such developments seem most urgent in the tropics, where the need to increase protein production by increasing grain legume yields is greatest, but they are also matters of concern in highly developed and intensive systems of agriculture in temperate zones, because of the need to conserve the very large amounts of fuel consumed in producing and applying nitrogenous fertiliser, often more than 10 000 kcal for each kilogram of nitrogen applied.

REFERENCES

1. ALLEN, O. N. and ALLEN, E. K. Plants in the subfamily *Caesalpinioideae* observed to be lacking nodules. *Soil Science*, **42**, 87–91 (1936)
2. ALMON, L. Concerning the reproduction of bacteroids, *Zentralblatt für Bakteriologie Parasitenkunde und Infektionskrankheiten*, **87**, 289–297 (1933)
3. ANDERSON, K. J. The effect of soil microorganisms on the plant–rhizobia association, *Phyton*, **8**, 59–73 (1957)
4. BARNET, Y. M. Bacteriophages of *Rhizobium trifolii*. 1. Morphology and host range, *Journal of General Virology*, **15**, 1–15 (1972)
5. BARNET, Y. M. and VINCENT, J. M. 'Self-lytic' strains of *Rhizobium trifolii, Australian Journal of Science*, **32**, 208 (1969)
6. BELL, F. and NUTMAN, P. S. Experiments on nitrogen fixation by nodulated lucerne, *Plant and Soil Special Volume*, 231–264 (1971)
7. BEZDICEK, D. F. Effect of soil factors on the distribution of *Rhizobium japonicum* sero-groups, *Soil Science Society of America Proceedings*, **36**, 305–307 (1972)
8. BHARDWAJ, K. K. R. Note on the growth of *Rhizobium* strains of dhaincha (*Sesbania cannabina*) in a saline-alkaline soil, *Indian Journal of Agricultural Science*, **42**, 432–433 (1972)
9. BHARDWAJ, K. K. R. and GAUR, A. C. Studies on the growth stimulating action of humic acid on bacteria, *Zentralblatt für Bakteriologie Parasitenkunde Infektionskrankheiten und Hygiene*, **126**, 649–699 (1971)
10. BONNIER, C. Anti-*Rhizobium* properties of extracts of certain *Leguminosae, Comptes Rendues de la Société Biologique*, **148**, 1894–1896 (1954)

127

Rhizobium *in the Soil*

11. BONNIER, C. La conservation du *Rhizobium* en sols stériles (*Rhizobium* spécifique des *Trifolium*), *Bulletin Institute of Agronomy Gembloux*, **23**, 359–367 (1955)

12. BROCKWELL, J. The presence of *Rhizobium meliloti* on pods of *Medicago tribuloides*, *Australian Journal of Experimental Agriculture and Animal Husbandry*, **5**, 141–143 (1965)

13. BURTON, J. C., ALLEN, O. N. and BERGER, K. C. The prevalence of strains of *Rhizobium phaseoli* in some Mid western soils, *Soil Science Society of America Proceedings*, **16**, 167–172 (1952)

14. CALDWELL, B. E. and HARTWIG, E. E. Serological distribution of soybean root-nodule bacteria in soils of South Eastern USA, *Agronomy Journal*, **62**, 621–622 (1970)

15. CHATEL, D. L. and PARKER, C. A. The colonization of host-root and soil by rhizobia. 1. Species and strain differences in the field, *Soil Biology and Biochemistry*, **5**, 425–432 (1973)

16. CHEN, M. and ALEXANDER, M. Survival of soil bacteria during prolonged desiccation, *Soil Biology and Biochemistry*, **5**, 213–221 (1973)

17. CHHONKAR, P. K. and SUBBA RAO, N. S. Fungi associated with legume root nodules and their effect on rhizobia, *Canadian Journal of Microbiology*, **12**, 1253–1261 (1966)

18. COOPER, R. Inoculation of legume seed by vacuum treatment, *Rothamsted Report for 1961*, 79 (1961)

19. DATE, R. A. Microbiological problems in the inoculation and nodulation of legumes, *Plant and Soil*, **32**, 703–725 (1970)

20. DAVEY, M. R. and COCKING, E. C. Uptake of bacteria by isolated higher plant protoplasts, *Nature*, **239**, 455–456 (1972)

21. DE ESCUDER, A. H. Q. A survey of rhizobia in farm soils at Wye College Kent, *Journal of Applied Bacteriology*, **35**, 109–118 (1972)

22. DELIN, S. Isolation of Thermoduric strains of *Rhizobium*, *Lantbrukshögskolans annaler*, Uppsala, **35**, 29–34 (1969)

23. DIATLOFF, A. Relationship of soil moisture, temperature and alkalinity to a soybean nodulation failure, *Queensland Journal of Agricultural and Animal Sciences*, **27**, 279–293 (1970)

24. DILWORTH, M. J. and PARKER, C. A. Development of nitrogen-fixing system in legumes, *Journal of Theoretical Biology*, **25**, 208–218 (1969)

25. DUDMAN, W. F. Antigenic analysis of *Rhizobium japonicum* by immunodiffusion, *Applied Microbiology*, **21**, 973–985 (1971)

26. ELKAN, G. H. Biochemical and genetical aspect of the taxonomy of *Rhizobium japonicum*, *Plant and Soil Special Volume*, 85–104 (1971)

27. FOTTRELL, P. F., O'CONNOR, S. and MASTERSON, C. L. Identification of the flavonol myricetin in legume seed and its toxicity to nodule bacteria, *Irish Journal of Agricultural Research*, **3**, 246–249 (1964)

28. GAUR, A. C. and MISRA, K. C. Effect of 2,4-dichlorophenoxyacetic acid on the growth of *Rhizobium* species *in vitro*, *Indian Journal of Microbiology*, **12**, 45–46 (1972)

29. GIBBINS, A. M. and GREGORY, K. F. Relatedness among *Rhizobium* and *Agrobacterium* species determined by three methods of nucleic acid hybridization, *Journal of Bacteriology*, **III**, 129–141 (1972)

30. GILTNER, W. and LONGWORTHY, H. V. Some factors influencing the longevity of soil microorganisms subjected to desiccation with special reference to soil solution, *Journal of Agricultural Research*, **5**, 927–942 (1916)

31. GODFREY, C. A. The carotenoid pigment and deoxyribonucleic acid base ratio of a *Rhizobium* which nodulates *Lotononis bainesii* Baker, *Journal of General Microbiology*, **72**, 399–402 (1972)

32. GRAHAM, P. H. The application of computer techniques to the taxonomy of the root nodule bacteria of legumes, *Journal of General Microbiology*, **35**, 511–517 (1964)

33. HAM, G. E., FREDERICK, L. R. and ANDERSON, I. C. Serogroups of *Rhizobium japonicum* in soybean nodules, *Agronomy Journal,* **63,** 69–72 (1971)

34. HAMDI, Y. A. Soil water tension and movement of rhizobia, *Soil Biology and Biochemistry,* **3,** 121–126 (1971)

35. HATTINGH, M. J. and LOUW, H. A. Clover rhizoplane bacteria antagonistic to *Rhizobium trifolii, Canadian Journal of Microbiology,* **15,** 361–364 (1969)

36. HEPPER, C. Genetics of red clover nodulation, *Rothamsted Report for 1972,* Part 1, 82 (1973)

37. HOLDING, A. J. and LOWE, J. F. Some effects of acidity and heavy metals on the *Rhizobium* in leguminous plant association, *Plant and Soil Special Volume,* 153–165 (1971)

38. JENSEN, H. L. Survival of *Rhizobium meliloti* in soil culture, *Nature,* **192,** 682–683 (1961)

39. JENSEN, H. L. The distribution of lucerne and clover rhizobia in agricultural soils in Denmark, *Tidsskrift for Planteavl,* **73,** 61–72 (1969)

40. JONES, D. G. and BURROWS, A. C. Acid production and symbiotic effectiveness in *Rhizobium trifolii, Soil Biology and Biochemistry,* **1,** 57–61 (1969)

41. JONES, D. G. and RUSSELL, P. E. The application of immunofluorescence techniques to host plant/nodule bacteria selectivity experiments using *Trifolium repens, Soil Biology and Biochemistry,* **4,** 277–282 (1972)

42. KECSKES, M. A survey of herbicide sensitivity and resistance of rhizobia, *Symposia Biologica Hungarica,* **11,** 454 (1970)

43. KECSKES, M. and MANNINGER, E. Effect of antibiotics on the growth of rhizobia, *Canadian Journal of Microbiology,* **8,** 157–159 (1962)

44. KLECZKOWSKA, J. Phage-induced changes in *Rhizobium, Rothamsted Report for 1958,* Part 1, 72 (1959)

45. KLECZKOWSKA, J. Establishment of phage and bacteria in a sterilised compost in a glasshouse. *Plant and Soil,* **37,** 425–429 (1972)

46. KOWALSKI, M. Lysogeny in *Rhizobium meliloti, Acta Microbiologica Polonica,* **15,** 119–128 (1966)

47. KRASSILNIKOV, N. A. and KORENIAKO, A. I. The influence of soil bacteria on the virulence and activity of *Rhizobium trifolii, Mikrobiologiya,* **13,** (i) 39 (1944)

48. MARSHALL, K. C. The nature of bacterium–clay interactions and its significance in survival of *Rhizobium* under arid conditions, *Proceedings IX International Congress of Soil Science, Adelaide,* Volume III, 275–280 (1968)

49. MARSHALL, K. C. Methods of study and ecological significance of rhizobium–clay interactions, in *Methods of Study in Soil Ecology* (edited by J. Phillipson), 107–110. Proceedings UNESCO/IBP Symposium, Paris (1970)

50. MASTERSON, C. L. Clover nodule bacteria survey, An Foras Talúntais Soils Division Research Department Dublin, pp. 78–79 (1961)

51. MEANS, U. M., JOHNSON, H. W. and ERDMAN, L. W. Competition between bacterial strains affecting nodulation in soybeans. *Soil Science Society of America Proceedings,* **25,** 105–108 (1961)

52. MOSTAFA, I. Y., FAKHR, I. M. I., BAHIG, M. R. E. and EL-ZAWAHRY, Y. A. Metabolism of organophosphorus insecticides. XIII. Degradation of malathion by *Rhizobium* species, *Archiv für Mikobiologie,* **86,** 221–224 (1972)

53. NUTMAN, P. S. The relation between nodule bacteria and the legume host in the rhizosphere and in the process of infection, in *Ecology of Soil-borne Plant Pathogens* (edited by K. F. Baker and W. C. Snyder), 231–247. University of California Press, Berkeley (1965)

129

Rhizobium *in the Soil*

54. NUTMAN, P. S. Microbiology of Broadbalk Soils; Legume nodule bacteria, *Rothamsted Report for 1968*, Part 2, 179–181 (1969)
55. NUTMAN, P. S. and READ, M. P. Symbiotic adaptation in local strains of red clover and nodule bacteria, *Plant and Soil*, **4**, 57–75 (1952)
56. NUTMAN, P. S. and ROSS, G. J. S. *Rhizobium* in the soils of the Rothamsted and Woburn farms, *Rothamsted Report for 1969*, Part 2, 148–167 (1970)
57. OBATON, M. Use of antibiotic-resistant spontaneous mutants for studying the ecology of *Rhizobium, Comptes Rendu Hebdomadaire des Séances de l'Academie des Sciences*, **272D**, 2630–2633 (1971)
58. PARKER, C. A. and GROVE, P. L. *Bdellovibrio bacteriovorus* parasitising *Rhizobium* in Western Australia, *Journal of Applied Bacteriology*, **33**, 253–255 (1970)
59. PATE, J. S. Perennial nodules on native legumes in the British Isles, *Nature*, **192**, 376–377 (1961)
60. PETERSON, H. B. and GOODING, T. H. The geographic distribution of *Azotobacter* and *Rhizobium meliloti* in Nebraska soils in relation to certain environmental factors, *University of Nebraska Agricultural Experimental Station Research Bulletin*, **121**, 1–24, (1941)
61. RAKE, A. V. Lack of DNA homology between the legume *Glycine max* and its symbiotic *Rhizobium* bacteria, *Genetics*, **71**, 19–24 (1972)
62. READ, M. P. The establishment of serologically identifiable strains of *Rhizobium trifolii* in field soils in competition with native microflora, *Journal of General Microbiology*, **9**, 1–14 (1953)
63. RICE, E. L. Inhibition of nodulation of inoculated legumes by leaf leachates from pioneer plant species from abandoned fields, *American Journal of Botany*, **58**, 368–371 (1971)
64. ROBINSON, A. C. Competition between effective and ineffective strains of *Rhizobium trifolii* in the nodulation of *Trifolium subterraneum*, *Australian Journal of Agricultural Research*, **20**, 827–841 (1969)
65. ROUGHLEY, R. J. The preparation and use of seed inoculants, *Plant and Soil*, **32**, 675–701 (1970)
66. ROVIRA, A. D. Plant root excretions in relation to the rhizosphere effect. II. A study of the properties of root exudate and its effect on the growth of microorganisms isolated from the rhizosphere and control soil, *Plant and Soil*, **7**, 195–208 (1956)
67. ROVIRA, A. D. *Rhizobium* numbers in the rhizospheres of red clover and paspalum in relation to soil treatment and numbers of bacteria and fungi, *Australian Journal of Agricultural Research*, **12**, 77–83 (1961)
68. SCHEFFLER, J. G. and LOUW, H. A. The serological characteristics of the clover rhizobia in the soils of the Stellenbosch district, *South African Journal of Agricultural Science*, **10**, 161–174 (1967)
69. SCHWINGHAMER, E. A. Studies on induced variation in the rhizobia. III. Host range modifications of *Rhizobium trifolii* by spontaneous and radiation-induced mutation, *American Journal of Botany*, **49**, 269–277 (1962)
70. SCHWINGHAMER, E. A. Antagonism between strains of *Rhizobium trifolii* in culture, *Soil Biology and Biochemistry*, **3**, 355–363 (1971)
71. SEN, A. and SEN, A. N. Survival of *Rhizobium japonicum* in stored air-dry soils, *Journals of the Indian Society of Soil Science*, **4**, 215–220 (1956)
72. SHERWOOD, M. T. Improved synthetic medium for the growth of *Rhizobium, Journal of Applied Bacteriology*, **33**, 708–713 (1971)
73. SKINNER, F. A. Influence of yeast extract on *Rhizobium trifolii, Rothamsted Report for 1972*, Part 1, 82–83 (1973)

74. SKRDLETA, V. Application of immunoprecipitation in agar gel for the serological typing of soybean root nodules, *Folia Microbiologica,* **14,** 32–35 (1969)
75. SINGH, B. N. Selection of bacterial food by soil flagellates and amoebae, *Annals of Applied Biology,* **29,** 18–22 (1942)
76. SINGH, B. N. Myxobacteria in soils and composts; their distribution, number and lytic action on bacteria, *Journal of General Microbiology,* **1,** 1–10 (1947)
77. SINGH, B. N. Studies on giant amoeboid organisms. 1. Distribution of *Leptomyxa reticulata* Goodey in soils of Great Britain and the effect of bacterial food on growth and cyst formation, *Journal of General Microbiology,* **2,** 8–14 (1948)
78. STENZ, E. Rhizobia in plant tissue cultures. Part 1. Investigations on the infection of plant cells by rhizobia *in vitro, Zentralblatt für Bakteriologie Parasitenkunde Infektionskrankheiten und Hygiene IIte Abteilung,* **126,** 142–150 (1971)
79. SUBBA RAO, N. S., LAKSHMI-KUMARI, M., SINGH, C. S. and MAGU, S. P. Nodulation of lucerne (*Medicago sativa*) under the influence of sodium chloride, *Indian Journal of Agricultural Science,* **42,** 384–386 (1972)
80. SULLIVAN, C. W. and CASIDA, L. E. Parasitism of *Azotobacter* and *Rhizobium* species by *Bdellovibrio bacteriovirus, Antonie van Leeuwenhoek Journal of Microbiology and Serology,* **34,** 188–196 (1968)
81. TRINICK, M. J. Symbiosis between *Rhizobium* and the non-legume, *Trema aspera, Nature,* **244,** 459–460 (1973)
82. TUZIMURA, K. and WATANABLE, I. The effect of rhizosphere of various plants on the growth of *Rhizobium, Soil Science and Plant Nutrition* (Tokyo), **8,** 13–17 (1962)
83. VAN SCHREVEN, D. A. Influence of seed disinfection with Amertan on rhizobia inoculation of *Medicago lupulina* L., *Plant and Soil,* **27,** 443–446 (1967)
84. VAN SCHREVEN, D. A. On the resistance of effectiveness of *Rhizobium trifolii* to a low pH, *Plant and Soil,* **37,** 49–55 (1972)
85. WEAVER, R. W., FREDERICK, L. R. and DUMENIL, L. C. Effect of soybean cropping and soil properties on numbers of *Rhizobium japonicum* in Iowa soils, *Soil Science,* **114,** 137–141 (1972)
86. WEINBERG, S. R. and STOCKY, G. Conjugation and genetic recombination of *Escherichia coli* in soil, *Soil Biology and Biochemistry,* **4,** 171–180 (1972)
87. WILKINS, J. The effects of high temperatures on certain root-nodule bacteria, *Australian Journal of Agricultural Research,* **18,** 299–304 (1967)
88. WILSON, J. K. Seasonal variation in the numbers of two species of *Rhizobium* in soil, *Soil Science,* **30,** 289–296 (1930)
89. WILSON, J. K. Relative numbers of two species of *Rhizobium* in soils, *Journal of Agricultural Research,* **43,** 261–266 (1931)
90. WILSON, J. R. and NORRIS, D. O. Some effects of salinity on *Glycine javanica* and its *Rhizobium* symbiosis, in *Proceedings of the XI International Grassland Congress, Paradise, Queensland, Australia,* 455–458 (1970)
91. ZEVENHUIZEN, L. P. T. M. Chemical composition of exopolysaccharides of *Rhizobium* and *Agrobacterium, Journal of General Microbiology,* **68,** 239–243 (1971)

7 Nitrification and Nitrifying Bacteria

N. WALKER

7.1 Introduction

Nitrification in soil is the process by which ammonia is oxidised to nitrate, and this oxidation takes place in two stages, mediated by two distinct groups of autotrophic bacteria. Because of its early association with Rothamsted, it seems appropriate to consider first the pioneer work of Robert Warington.[30-33] During the late 1880s and early 1890s Warington studied the problems of nitrate formation and losses from soils just prior to the epoch-making researches of Winogradsky and, although not a trained bacteriologist, he made several important contributions to the subject, establishing some basic facts about nitrification and almost succeeding in the isolation of the responsible bacteria. Warington[30] confirmed the work of Schloesing and Muntz,[21] who had shown that nitrification in sewage could be inhibited by chloroform, by his demonstration that antiseptics (for example, chloroform, carbon disulphide, phenol) would stop nitrification in soil. Re-inoculation with fresh soil or with enrichment cultures was necessary to start the nitrifying process again. Ammonia was nitrified best in soil at about 30°C, at a slightly alkaline reaction and in presence of a base such as calcium carbonate. Light was inhibitory to nitrification. Warington[31, 32] further established that the process took place in two stages—conversion of ammonia to nitrite and conversion of nitrite to nitrate—and that these oxidations were effected by different species of bacteria. Although it is likely that at times he had virtually pure cultures of the responsible bacteria, he failed to recognise unequivocally their autotrophic nature.[32]

At about that time Winogradsky[41, 42] published his signal contributions to our knowledge of the nitrifying bacteria and, in

133

particular, he established that the two types of nitrifying organisms were true autotrophs deriving their carbon solely from carbon dioxide and their energy from the oxidation of either ammonia or nitrite, respectively. Winogradsky made important advances in methodology as a direct consequence of his appreciation of the autotrophic nature of the organisms. His earlier studies on sulphur- and iron-oxidising micro-organisms provided relevant experience on which to base his views of the similar autotrophic characteristics of the nitrifiers. By using a purely inorganic medium and even specially purifying a specimen of ammonium sulphate, he was able finally to obtain pure cultures and to establish their autotrophic nature. He discovered a variety of ammonia- and nitrite-oxidising bacteria which differed morphologically and in growth characteristics. Some 40 years later, in collaboration with Helene Winogradsky, Winogradsky[43] surveyed the available information about nitrifying organisms and, after reviewing the work of many others, declared that the knowledge of these organisms had made little progress since his own early studies. He regretted that no new methods of study had emerged which might have been useful in assessing their role in Nature or in examining the problem of the nitrifying power of soils. The several pure strains he had obtained were assigned to three genera: *Nitrosomonas,* with a pH optimum of between 8.6 and 9.2 and characteristic of well-manured soils; *Nitrosocystis,* with a pH optimum of between 7.4 and 7.8 and typical of forest soils; and *Nitrosospira,* with a pH optimum of between 7.0 and 7.2, strains of which were found only in poor, virgin soils. These different genera of ammonia-oxidisers could be differentiated on silica gel plates enamelled with a surface layer of chalk, basic magnesium carbonate or magnesium ammonium phosphate, so giving a medium completely devoid of organic substances. None of the species grew at pH values below 6, and he ascribed the greatest activity to *Nitrosomonas* spp., less to *Nitrosocystis* spp. and least to *Nitrosospira* spp. He thought that abundance of nitrosomonads indicated an actively nitrifying regime. In arable soils the numbers of nitrifiers usually found were some thousands per gram of soil and the only situation where nitrifiers reached formidable numbers was in activated sludge, where the population could be reckoned in millions per gram. Nitrite-oxidising bacteria were recognised by their growth on inorganic silica gel medium enamelled with a mixture of chalk and kaolin to give a porcelain-like surface on which their tiny brownish yellow colonies were fairly readily observed. In contrast, the colonies formed by the ammonia-oxidisers were

134

surrounded by clear zones caused by dissolution of the chalk by the nitrous acid produced. Nitrite-oxidisers were inhibited by moderate amounts of ammonia, especially at alkaline reactions. The Winogradskys also considered faulty techniques to be the prime reason for occasional reports of heterotrophic nitrifiers.

Now, after a second 40 year period in which, admittedly, much more work has been devoted to the autotrophic nitrifying bacteria, it is still true to say that relatively little more has been added to our knowledge until very recently. Moreover, quite a number of investigations have been repetitive and have added little to the information that was already known to the Winogradskys.

7.2 Isolation and cultural methods

With recognition of the autotrophic nature of the nitrifiers and increasing knowledge of their physiology, the methods for isolating and growing these micro-organisms require a purely inorganic medium, and it was by his use of silica gel that Winogradsky was able to prepare media free from organic substances and thus to minimise contamination by heterotrophic contaminants. Ammonia oxidisers require a medium with small amounts of magnesium, calcium and a phosphate and a trace quantity of iron. Sodium chloride is sometimes added but is not essential, except for marine nitrifiers, although it is not always an adequate substitute for sea-water. The need for traces of copper is doubtful. Nitrite oxidisers require a trace of molybdenum.[6, 44] The slow growth rate of nitrifiers, perhaps due to the low energy yield from the ammonia or nitrite oxidations as compared with oxidations of carbon compounds, probably also contributes to the low yield of cells. Maintenance of a suitable pH value and adequate aeration are the other main growth requirements. In starting an enrichment culture it is an advantage to use only moderate concentrations of ammonium salts or of nitrite (preferably less than 0.1%) but more substrate can be added as growth proceeds. Winogradsky selected sparingly soluble buffers such as calcium carbonate, magnesium ammonium phosphate or basic magnesium carbonate because an excess of these could be used, whereas with soluble carbonates, e.g. sodium carbonate, an excess must be avoided if the pH value is to be kept below about 8.5. When such insoluble buffers are used in liquid media, the cells mostly become adsorbed on the particles and the supernatant solution may appear clear. However, this should not be taken to imply that adsorption on solid particles is at all necessary for the growth of nitrifiers, as is sometimes assumed. Nitrifying bacteria will grow equally well in media free of insoluble particles, such as chalk.

When ammonia-oxidising bacteria are grown on silica gel plates coated with a layer of chalk or magnesium carbonate, the insoluble carbonates are dissolved by the nitrous acid produced and so the tiny colonies, surrounded by clear zones, may be recognised and counted. Many workers have used this technique since its introduction by Winogradsky, and a slightly modified version of it was recently described in detail by Soriano and Walker.[27] The technique is suitable for counting ammonia-oxidising nitrifiers, and comparisons with the alternative dilution method for determining most probable numbers have been reported several times (see, for example, references 9 and 22). The plating method has the advantage that, after counting, organisms may be isolated by colony picking.[26, 27] Pure cultures of nitrifying bacteria may be obtained by such a plating procedure or by dilution of an enrichment culture until the volume to be used as an inoculum contains only one or two cells; a large number of tubes are then inoculated. After suitable incubation, usually for 2–3 weeks, some of these tubes will contain pure cultures. In any case, the methods are tedious because of the slow growth rate of the organisms and the need to check the purity of the cultures obtained, but the colony-picking method is sometimes quicker and often more satisfactory. Originally, Winogradsky isolated pure cultures by plating an enrichment culture on to silica gel medium and making transfers from parts of the silica gel apparently devoid of colonies. This was the 'negative plating' technique. Nitrite-oxidisers may be isolated in pure culture best by plating a suitable enrichment culture on to a mineral salts, nitrite, agar plate made from washed and purified agar and, after incubation, likely colonies can be picked off with a micropipette. Media containing certain antibiotics to inhibit heterotrophic contaminants have also been used.[8]

7.3 Classification of nitrifying bacteria

Winogradsky described no fewer than six genera of nitrifying bacteria—three of ammonia-oxiders and three of nitrite-oxidisers—and until recently these have remained recognised, although, except for *Nitrosomonas* and *Nitrobacter,* few workers have studied the other species. In the seventh edition of *Bergey's Manual*[2] the autotrophic nitrifying bacteria are included in the family *Pseudomonadaceae*, which is unfortunate because *Nitrosomonas europaea* is probably the only species which can be considered to fit the description of a pseudomonad. (The classification of this whole group of organisms is currently being revised for the next edition of the *Manual;* see reference 36.) In recent years new species of nitrifiers have been isolated, especially by Watson and his colleagues at Woods Hole Oceanographic Institute; several species originally described by Winogradsky have been re-isolated, and so the validity of certain genera of these chemolithotrophs is now established without doubt. The tendency of some workers (e.g. Bissett and Grace[3]) to dismiss all genera except *Nitrosomonas* and *Nitrobacter* spp. as invalid must therefore be discounted. Without referring in detail to every reported isolation of nitrifiers, an attempt will now be made to summarise present knowledge of the taxonomy of the recognised genera and species.

Ammonia-oxidisers

Five genera of ammonia-oxidisers have been reported by the Winogradskys[43]—namely, *Nitrosomonas, Nitrosocystis, Nitrosococcus, Nitrosospira* and *Nitrosogloea*. In agreement with Watson (personal communication), the latter, *Nitrosogloea*, of which Helene Winogradsky[40] detailed three species—*N. merismoides, N. schizobacteroides* and *N. membranacea*—all isolated from activated sludge, must be considered doubtful and their names rejected. This is because zoogloea-formation cannot be regarded as a reliable, stable character and furthermore the above three species were not described sufficiently well for them to be identified again beyond doubt. It is not always clear what is meant by cyst-formation, a property that presumably should characterise the genera *Nitrosocystis* or *Nitrocystis*. The formation of aggregates of cells, sometimes enclosed in some kind of sheath or coat, has been observed in various cultures, often when impure. Consequently, there is some doubt regarding the validity of this name. Nevertheless, the name *Nitrosocystis coccoides* still seems to be valid. A typical strain of this species was isolated from forest soil by Romell,[20] working in Winogradsky's laboratory, and described as forming hard bodies up to about 0.5 mm in diameter. There is little doubt that *Nitrosocystis javanensis*, described by Winogradsky, should be regarded as a *Nitrosomonas* sp. and this leaves *Nitrosocystis coccoides* as the only valid species of that genus. Some confusion was caused by Winogradsky, who reclassified his original Zürich strain of *Nitrosomonas europaea* as a *Nitrosocystis* sp., and then first by Imsenecki and later Grace, who considered that Romell's culture, mainly on indirect evidence, was contaminated by a myxobacterial species. Soriano and Walker[27] have isolated a number of *Nitrosocystis coccoides* strains from soil. Morphologically they are distinct from *Nitrosomonas* because they have an irregular stumpy rod-like shape and a jerky motility, and form hard colonies on silica gel or on washed agar plates; these colonies often can be pushed along the surface of the plate with a micropipette without fragmenting. One strain, isolated some 10 years ago from soil taken from a Park Grass plot, has been found identical in morphology, fine structure and DNA base ratio (%G + C content) with a strain isolated by Watson *et al.*[39] Watson has proposed renaming this organism *Nitrosolobus multiformis* on account of its peculiar morphology, and this proposal is still *sub judice*. Previously Watson[34] isolated a marine nitrifying organism which he designated as *Nitrosocystis oceanus* nov. sp., but later he has reconsidered this and suggested that, as it is really a large coccus, it would be better to rename it *Nitrosococcus oceanus*. Consequently, he would retain the genus *Nitrosococcus* to include all spherical nitrifying organisms having a diameter of 1.5 μm or greater. Furthermore, he proposed that *Nitrosocystis coccoides* should be treated as a synonym of *Nitrosococcus nitrosus*. This seems hardly justified in view of the differences between *Nitrosocystis coccoides* (or *Nitrosolobus multiformis* as proposed by Watson *et al.*) and *Nitrosococcus nitrosus,* which has been isolated since Winogradsky's time both by Sims and Collins[22] and by Soriano.[25] Colonies of *N. nitrosus* on washed agar plates are smooth, circular soft colonies which readily disintegrate when pierced by a micropipette, the contents of the colony being sucked up into the pipette by capillary action. This is in marked contrast to the behaviour of colonies of *Nitrosocystis coccoides*. There are also differences in the %G + C content of their DNA. Calculated from determinations of the T_m ('melting point') of DNA preparations, the Park Grass strain of *Nitrosocystis coccoides* had DNA with $55.2 \pm 0.5\%$ G + C, whereas the DNA from Soriano's strain of *Nitrosococcus nitrosus* contained 45.4% G + C. The growth rates of *N. nitrosus* and *Nitrosomonas europaea* are very similar, both of them having a doubling time of about 11–12 h at 25°, but the former is a larger organism, as shown by dry weight determinations of the cultures. *Nitrosomonas* is the genus for rod-shaped organisms that oxidise ammonia to nitrite, and two species are listed in *Bergey's Manual*[2]—*N.*

europaea and *N. monocella*—but, in agreement with Watson, these two species are similar enough to be regarded as the same. *N. javanensis* may well be synonymous with *N. europaea,* and there does not seem to be any justification for treating *N. oligocarbogenes*[5] as a distinct species merely on the basis of its differing ratio of 70 : 1 of nitrogen oxidised to carbon assimilated compared with 35 : 1 of N oxidised/C assimilated for *N. europaea.* A *Nitrosomonas* strain that I isolated from activated sludge a few years ago proved to be identical in its DNA % G + C content and in fine structure with a strain isolated by Watson from Chicago sewage. These isolates are slightly larger than typical *N. europaea* strains and have prominent connections between dividing cells. Under the electron microscope two or sometimes more cells can be seen to be interconnected with no separating cross-walls being visible. These morphological differences may possibly justify creating a separate species for these strains. Finally, there is the genus *Nitrosospira* which Winogradsky designated to include two strains—*N. briensis* and *N. antarctica.* Watson,[35] who has recently reported the isolation of a *Nitrosospira* sp., rightly regards the two species as identical and proposes using the species *Nitrosospira briensis* for the type species. *Nitrosospira* species have also been isolated by Soriano, by Palleroni and by Walker several years ago. Moreover,

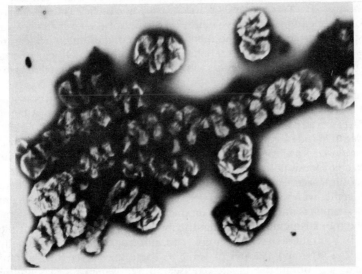

Figure 7.1 Electron micrograph of a Nitrosospira *sp. isolated from a Spitzbergen soil. Negatively stained with uranyl acetate; magnification × 25 000; reduced by two-thirds on reproduction*

there seems to be evidence that, far from being dismissed as of doubtful identity, *Nitrosospira* sp. may occur widely in arable or virgin soils (Soriano and Walker).[27] The spiral nature of these species is not obvious under the phase contrast microscope but is readily recognised under the electron microscope (*Figure 7.1*). The organisms do not possess the well-developed cell membrane system characteristic of other ammonia oxidisers.

Nitrite-oxidisers

Organisms of the genus *Nitrobacter* (Winogradsky) are short rods, 0.6–0.8 μm wide and 1.0–1.2 μm long; often pear- or wedge-shaped; Gram-negative; and reproducing by budding.[45] *Nitrobacter winogradskyi* is now recognised as the type species. Nelson[17] described a motile species of *Nitrobacter* which had a long polar flagellum and he regarded it as a new species, *N. agilis,* but it very probably is synonymous with *N. winogradskyi.* Motility is often not readily observed in these organisms and its presence or absence is not a sufficient reason for creating a new species. Of some 20 pure cultures of *Nitrobacter* spp. isolated from Rothamsted soils, Soriano and Walker[27] have observed motility in all of them at some stage or other in their growth. When grown in liquid cultures, many *Nitrobacter* strains, after several days' growth, form a fine pellicle on the surface of the medium, and this pellicle is usually so water-repellent as to spread up the sides of the tube. The cells contain lipids[4] which are probably responsible for the water-repellency. Other micro-organisms (e.g. *Nocardia* spp.), also rich in lipids, exhibit the same phenomenon. Recently, Watson and Waterbury[37] isolated two new, morphologically interesting, nitrite-oxidising bacteria from sea-water. *Nitrospina gracilis nov. gen. nov. sp.* was isolated from Atlantic Ocean water and found to be a long, slender, non-motile, Gram-negative rod; a typical cytomembrane system was lacking. *Nitrococcus mobilis nov. gen. nov. sp.* was isolated from sea-water collected in the Pacific Ocean near the Galapagos Islands. The organism was a large motile coccus growing singly, in pairs or in small aggregates held together with slime. The cells were Gram-negative, about 1.5–1.8 μm in diameter, and possessed a unique tubular cytomembrane system, the tubular membranes being distributed randomly throughout the cell. Both these organisms were strict autotrophs growing optimally at 25–30°C and at a pH of 7.5–8.0. These species seem to be rare and the dominant nitrite-oxidiser in sea-water appears to be *Nitrobacter.*

7.4 Biochemistry of nitrification

As yet, our knowledge of the biochemistry of the nitrifying bacteria is limited. Except for *Nitrobacter winogradskyi,* these bacteria are all strict autotrophs, obtaining their carbon from CO_2 and energy from the oxidation of ammonia or of nitrite, respectively. Nitrifying bacteria thus synthesise all their organic components, including proteins, enzymes, nucleic acids, cytochrome pigments, etc., from CO_2. The first biochemical studies were by Meyerhof,[15] who determined the optimum pH for ammonia oxidation by washed *Nitrosomonas* cells and for nitrite oxidation by washed *Nitrobacter* cells. He also examined the effect on *Nitrosomonas* and *Nitrobacter* cells of various organic compounds, of which some were highly inhibitory, e.g. narcotics, also amines such as guanidine, aminoguanidine, aniline, *p*-nitroaniline and *p*-nitroso-dimethylaniline. Bömeke[5] found that washed *Nitrosomonas* cells showed a small but measurable rate of O_2 uptake, i.e. a resting or endogenous respiration. This is easily confirmed, given adequate amounts of cells, which may best be grown in a clear, precipitate-free medium (cf. reference 23). Nitrifiers, being strict aerobes, are also inhibited by appropriate agents that interfere with cytochrome oxidase or catalase, e.g. cyanide, cyanate, azide, chlorate and some chelating agents. Thus, although their prime source of energy is obtained from an inorganic oxidation, the cells themselves possess a metabolism comparable with that of heterotrophic organisms.[29] More modern work has shown that CO_2 is assimilated through the Calvin cycle which operates in plants and in other autotrophic microorganisms and where the necessary reducing power for reducing CO_2 is derived from reduced di- and tri-phospho-pyridine nucleotides. The primary energy reaction of the nitrifiers is, of course, quite specific to these organisms (although certain methane-oxidising bacteria have been shown to oxidise ammonia to nitrite when grown in presence of CH_4, CO_2 and O_2; see reference 11).

The pathway of ammonia oxidation can be summarised by saying that the first intermediate seems to be hydroxylamine, even though there are certain differences in the stability of the ammonia and hydroxylamine-oxidising systems of *Nitrosomonas* cells. Hydroxylamine is oxidised to nitrite only when its concentration does not exceed about M/3000. Theoretically there could be at least one more oxidation stage between hydroxylamine and nitrite but the identity of this is not yet established. The possible nature of such intermediates is discussed in detail by Lees.[13] Kluyver and Donker (see reference 13) first formulated, in 1926, a stepwise oxidation of ammonia as follows:

$$NH_3 \rightarrow NH_2OH \rightarrow H_2N_2O_2 \quad \text{or} \quad \rightarrow HNO_2$$
$$NH(OH)_2$$

The identity of the intermediate between hydroxylamine and nitrite is still a matter for speculation, as evidence for the occurrence of hyponitrite is not convincing and dihydroxy-ammonia is too unstable a substance to be prepared. Attempts to isolate and purify an ammonia-oxidising enzyme system have not yet been successful. Various workers have obtained cell-free preparations with which the production of very small amounts of nitrite from ammonia have been demonstrated, but these have yielded little or no information on the nature of the enzyme system involved. When *Nitrosomonas* cells are disintegrated ultrasonically or by physical rupture, the ammonia-oxidising capacity is lost but some hydroxylamine-oxidising function remains, although only a part of the oxidised hydroxylamine is accounted for as nitrite. Rees[18, 19] has partially purified the hydroxylamine oxidase from *Nitrosomonas europaea* and studied some of its properties. Watson, Asbell and Valois[38] have also demonstrated the oxidation of ammonia to nitrite by cell-free preparations of *Nitrosocystis oceanus* when the cells were ruptured in presence of buffer, ATP, Mg^{2+} ions and phosphate. In recent years more work has been reported on the organic metabolism of autotrophic nitrifiers, especially on the functioning of the reductive pentose phosphate cycle involved in CO_2 assimilation (see review by Kelly).[12]

Little is yet known of the nitrite-oxidising enzyme system, although active cell-free preparations have been reported (see, for example, reference 1). On the other hand. *Nitrobacter* spp. are capable of heterotrophic metabolism and growth even though growth is slower on an organic carbon source such as acetate or formate. With a mixture of nitrite and formate, *Nitrobacter* grows more quickly than on formate alone but nevertheless more slowly than on nitrite alone. Acetate and certain amino acids can be assimilated by *Nitrobacter* cells, as shown by experiments with ^{14}C-labelled compounds, so that *Nitrobacter* spp. must be considered as facultative autotrophs (see reference 24). *Nitrosocystis* spp. possess urease activity and so can grow on urea, a property not shared by *Nitrosomonas*. Growth on formate may well be a stage in the evolution of autotrophy as the utilisation of one-carbon compounds is a development from complete reliance on CO_2.

Both the ammonia- and the nitrite-oxidising autotrophs are strictly aerobic organisms and depend on cytochrome systems[14] for electron transport ultimately to oxygen. Cells of the two kinds of nitrifiers are rich in a, b, c and o type cytochromes, which give the deep red colour to pellets of centrifuged washed cells. Much cytochrome material is readily released when thick cell suspensions are subjected to alternate freezing and thawing.[23]

It is not possible in the space available to refer to all the recent publications dealing with chemolithotrophic biochemistry, but there is clearly much scope for further detailed study, especially on the nature of the ammonia- and nitrite-oxidising enzymes.

7.5 Nitrification in soil

In spite of various reports of the nitrifying ability of heterotrophic organisms, including fungi, there is little doubt that nitrification in soils is predominantly effected by the autotrophic nitrifying bacteria. Provided that normal conditions for microbial activities obtain (that is to say, adequate moisture, suitable aeration conditions, favourable pH and a source of nutrients), then autotrophic nitrifying bacteria are ubiquitous wherever ammonia and nitrite are available. Numbers of nitrifiers in most soils are never very high—usually some thousands per gram in a fertile arable soil—probably because of their slow growth rate[16] and their requirement for considerable amounts of ammonia or nitrite as energy sources.[27] Given favourable conditions, numbers are largely dependent on the amount of ammonia present and so numbers in sewage beds or in activated sludge may be much higher than in soils. Nitrates are easily leached from soil, whereas ammonia is better adsorbed and retained on the surfaces of clay particles. Consequently, attempts have been made to inhibit nitrification and so prevent losses of fertiliser nitrogen from soil. Goring[7] reported the use of the inhibitor N-serve (2-chloro-6-(trichloromethyl)-pyridine) as a means of controlling nitrification in soil. This is an effective agent for a certain time but in due course the substance itself may be degraded by soil micro-organisms and thus become ineffective. The best use of such agents requires careful consideration of all the factors involved, including the time of year, soil properties, rainfall, crop, etc. From time to time there have been suggestions that some plants, particularly grasses, may inhibit nitrifiers in soil near their roots. Up to now there is no convincing evidence for the excretion of any specific toxic substance from the roots of such plants. It is more probable that the dense network of fine roots which characterises many grasses serves to interfere with access of oxygen to the soil and so decreases the aeration capacity needed for the nitrifiers.

Conditions that are favourable to nitrifying bacteria are also largely those that favour plant growth, especially conditions of good aeration, neutral pH and nitrogen supply. This has occasionally led to suggestions that the presence of an active nitrifying regime in soil may be a useful indicator of soil fertility, but chemical methods for assessing soil fertility are probably preferable. Ecological studies of the distribution of nitrifying bacteria have at times indicated a small degree of adaptation on the part of nitrifiers to varying soil conditions, especially as regards pH optima (see, for example, reference 28). However, our knowledge of nitrifier ecology is still very fragmentary, particularly with regard to the habitats and distribution of the different species and genera, and further work on these interesting organisms is desirable.

Acknowledgement

I am grateful to Mrs. Pamela Evans of the Plant Pathology Department for kindly preparing the photograph of a *Nitrosospira* sp.

REFERENCES

1. ALEEM, M. I. H. and ALEXANDER, M. Cell-free nitrification by *Nitrobacter, Journal of Bacteriology,* **76,** 510–514 (1958)
2. *Bergey's Manual of Determinative Bacteriology,* 7th edition. Bailliere, Tindall and Cox, London (1957)
3. BISSETT, K. A. and GRACE, JOYCE B. The nature and relationships of autotrophic bacteria, in *Autotrophic Micro-organisms:* 4th Symposium of the Society for General Microbiology. Cambridge University Press (1954)
4. BLUMER, M., CHASE, T. and WATSON, S. W. Fatty acids in the lipids of marine and terrestrial nitrifying bacteria, *Journal of Bacteriology,* **99,** 366–370 (1969)
5. BÖMEKE, H. Beiträge zur Physiologie nitrifizierender Bakterien, *Archiv für Mikrobiologie,* **10,** 385–445 (1939)
6. FINSTEIN, M. S. and DELWICHE, C. C. Molybdenum as a micronutrient for *Nitrobacter, Journal of Bacteriology,* **89,** 123–128 (1965)
7. GORING, C. A. Control of nitrification of ammonium fertilisers and urea by 2-chloro-6-(trichloromethyl)-pyridine, *Soil Science,* **93,** 431–439 (1967)
8. GOULD, G. W. and LEES, H. The isolation and culture of the nitrifying organisms. Part I. *Nitrobacter, Canadian Journal of Microbiology,* **6,** 299–307 (1960)

9. HÖFLICH, GISELA. Möglichkeiten zur quantitativen Ermittlung der Nitrificationsbakterien, *Zentralblatt für Bakteriologie Parasitenkunde, Infektionskrankheiten und Hygiene IIte Abteilung,* **123,** 138–146 (1969)

10. HOFMAN, T. and LEES, H. Biochemistry of the nitrifying organisms. 4. The respiration and intermediary metabolism of *Nitrosomonas, Biochemical Journal,* **54,** 579–583 (1953)

11. HUTTON, W. E. and ZOBELL, C. E. Production of nitrite from ammonia by methane oxidizing bacteria, *Journal of Bacteriology,* **65,** 216–219 (1953)

12. KELLY, D. P. Autotrophy: concepts in lithotropic bacteria and their organic metabolism, *Annual Review of Microbiology,* **25,** 177–210 (1971)

13. LEES, H. The biochemistry of the nitrifying bacteria, in *Autotrophic Micro-organisms,* 84–98. Cambridge University Press (1954)

14. LOZINOV, A. B. and ERMACHENKO, V. A. The physiological role of cytochrome in nitrite bacteria, *Microbiology* [English translation of *Mikrobiologia*], **31,** 788–793 (1963)

15. MEYERHOF, O. Untersuchungen über den Atmungsvorgang nitrifizierender Bakterien. III. Die Atmung des Nitritbildners und ihre Beeinflussung durch chemischen Substanzen, *Pflügers Archiv für Physiologie,* **166,** 240–242 (1917)

16. MORILL, L. G. and DAWSON, J. E. Growth rates of nitrifying chemoautotrophs in soil, *Journal of Bacteriology,* **83,** 205–206 (1962)

17. NELSON, D. H. Isolation and characterization of *Nitrosomonas* and *Nitrobacter, Zentralblatt für Bakteriologie, Parasitenkunde und Infektionskrankheiten, IIte Abteilung,* **83,** 280–311 (1931)

18. REES, M. K. Studies of the hydroxylamine metabolism of *Nitrosomonas europaea.* I. Purification of hydroxylamine oxidase, *Biochemistry,* **7,** 353–366 (1968)

19. REES, M. K. Studies of the hydroxylamine metabolism of *Nitrosomonas europaea.* II. Molecular properties of the electron-transport particle, hydroxylamine oxidase, *Biochemistry,* **7,** 366–372 (1968)

20. ROMELL, L. G. En nitrikbakterie ur Svensk Skogsmark, *Meddelanden fran Statens Skogsförsöksanstalt* (Stockholm), **24,** 57–66 (1928)

21. SCHLOESING, T. and MUNTZ, A. Sur la nitrification par les ferments organisés, *Comptes Rendues de l'Academie des Sciences* (Paris), **84,** 301–303 (1877)

22. SIMS, C. M. and COLLINS, F. M. The numbers and distribution of ammonia-oxidizing bacteria in some Northern Territory and South Australian soils, *Australian Journal of Agricultural Research,* **11,** 505–512 (1960)

23. SKINNER, F. A. and WALKER, N. Growth of *Nitrosomonas europaea* in batch and continuous culture, *Archiv für Mikrobiologie,* **38,** 339–349 (1961)

24. SMITH, A. J. and HOARE, D. S. Acetate assimilation by *Nitrobacter agilis* in relation to its 'obligate autotrophy', *Journal of Bacteriology,* **95,** 844–855 (1968)

25. SORIANO, S. Re-isolation of the *Nitrosococcus* of Winogradsky, *Annales de l'Institut Pasteur* (Paris), **105,** 349–352 (1963)

26. SORIANO, S. and WALKER, N. Isolation of ammonia-oxidizing autotrophic bacteria, *Journal of Applied Bacteriology,* **31,** 493–497 (1968)

27. SORIANO, S. and WALKER, N. The nitrifying bacteria in soils from Rothamsted classical fields and elsewhere, *Journal of Applied Bacteriology,* **36,** 523–529 (1973)

28. UL'YANOVA, O. M. Adaptation of *Nitrosomonas* to existence conditions on various natural substrates, *Microbiology* [English translation of *Mikrobiologiya*], **30,** 216–220 (1961)

29. WALLACE, W., KNOWLES, S. E. and NICHOLAS, D. J. D. Intermediary metabolism of carbon compounds by nitrifying bacteria, *Archiv für Mikrobiologie,* **70,** 26–42 (1970)

30. WARINGTON, R. On nitrification, *Journal of the Chemical Society,* **33,** 44–51 (1878)

Nitrification and Nitrifying Bacteria

31. WARINGTON, R. On nitrification, Part II, *Journal of the Chemical Society*, **35**, 429–456 (1879)

32. WARINGTON, R. On nitrification, Part III, *Journal of the Chemical Society*, **45**, 637–672 (1884)

33. WARINGTON, R. On nitrification, Part IV, *Journal of the Chemical Society*, **59**, 484–529 (1891)

34. WATSON, S. W. Characteristics of a marine nitrifying bacterium, *Nitrosocystis oceanus* nov. sp., *Limnology and Oceanography*, **10** (Suppl.), 274–289 (1965)

35. WATSON, S. W. Re-isolation of *Nitrosospira briensis* S. Winogradsky and H. Winogradsky 1933, *Archiv für Mikrobiolgie*, **75**, 179–188 (1971)

36. WATSON, S. W. Taxonomic considerations of the family *Nitrobacteraceae* Buchanan, *International Journal of Systematic Bacteriology*, **21**, 254–270 (1971)

37. WATSON, S. W. and WATERBURY, J. B. Characteristics of two marine nitrite-oxidising bacteria, *Nitrospina gracilis nov. gen. nov. sp.* and *Nitrococcus mobilis nov. gen. nov. sp.*, *Archiv für Mikrobiologie*, **77**, 203–230 (1971)

38. WATSON, S. W., ASBELL, MARY ANN and VALOIS, FREDERICA W. Ammonia oxidation by cell-free extracts of *Nitrosocystis oceanus*, *Biochemical and Biophysical Research Communications*, **38**, 1113–1119 (1970)

39. WATSON, S. W., GRAHAM, LINDA B., REMSEN, C. C. and VALOIS, FREDERICA W. A lobular, ammonia-oxidising bacterium, *Nitrosolobus multiformis nov. gen. nov. sp.*, *Archiv für Mikrobiologie*, **76**, 183–203 (1971)

40. WINOGRADSKY, HELENE. Contribution a l'étude de la microflore nitrificatrice des boues activées de Paris, *Annales de l'Institut Pasteur*, **58**, 326–340 (1937)

41. WINOGRADSKY, S. Recherches sur les organismes de la nitrification, *Annales de l'Institut Pasteur* (Paris), **4**, 213–231 (1890)

42. WINOGRADSKY, S. Recherches sur les organismes de la nitrification, 2^c Memoire, *Annales de l'Institut Pasteur* (Paris), **4**, 257–275 (1890)

43. WINOGRADSKY, S. and WINOGRADSKY, HELENE. Nouvelles recherches sur les organismes de la nitrification, *Annales de l'Institut Pasteur* (Paris), **50**, 350–432 (1933)

44. ZAVARZIN, G. A. The participation of molybdenum in the oxidation of nitrites by nitrifying bacteria, *Doklady Akademii Nauk SSSR*, **113**, 1361–1362 (1957) [in Russian]

45. ZAVARZIN, G. A. and LEGUNKOVA, R. The morphology of *Nitrobacter winogradsky*, *Journal of General Microbiology*, **21**, 186–190 (1959)

Suggestions for further reading

Autotrophic Micro-organisms: 4th Symposium of the Society for General Microbiology (edited by B. A. Fry and J. L. Peel). Cambridge University Press (1954)

LEES, H. *Biochemistry of Autotrophic Bacteria.* Butterworths, London (1955)

ENGEL, H. Die Nitrifikanten, in *Handbuch der Pflanzenphysiologie* (edited by W. Ruhland) Volume 5/2, 664–681. Springer-Verlag, Berlin (1960)

MEIKLEJOHN, J. The nitrifying bacteria: a review, *Journal of Soil Science*, **4**, 59–68 (1953)

LARSEN, HELGE. Chemosynthesis, in *Handbuch der Pflanzenphysiologie* (edited by W. Ruhland) Volume 5/2, 613–648. Springer-Verlag, Berlin (1960)

PECK, H. D. JR. Energy-coupling mechanisms in chemolithotrophic bacteria, *Annual Review of Microbiology*, **22**, 489–518 (1968)

WATSON, S. W. Taxonomic considerations of the family *Nitrobacteraceae* Buchanan, *International Journal of Systematic Bacteriology*, **21**, 254–270 (1971)

146

8 Soil Protozoa—Animalcules of the Subterranean Microenvironment

J. F. DARBYSHIRE

8.1 The development of soil protozoology

Although many nineteenth century biologists detected protozoa in soils, it is only in the present century that protozoa have been widely accepted as active components of the soil community. The numerous experiments of Russell and his colleagues between 1905 and 1913, using soil treated with heat or chemicals such as toluene or chloroform, provided the stimulus for this change of outlook. Russell[59] was concerned originally with relationships between soil fertility and the absorption of oxygen by soil. He found that both phenomena were greatly diminished if the soil had been heated previously to 120°C. On one occasion when the temperature only rose to 95°C, because the steriliser was defective, the rate of oxidation was considerably increased in comparison with untreated soil.[17] This unexpected result led to a series of experiments, which had a considerable impact on practical horticulture as well as soil protozoology. Plant yields and the amount of oxidation showed a marked increase after these so-called partial sterilisation treatments. Russell and Hutchinson[60, 61] concluded that this improvement in soil fertility was due mainly to the removal of some unfavourable biological factor. They suggested that this factor was

147

the bacteriophagous soil protozoan population. Although modern interpretations of the partial sterilisation effect (see Chapter 11) suggest that this explanation is incorrect, their hypothesis was controversial enough to create renewed interest in these animals. This hypothesis presupposed that trophic protozoa were present in soil and that there was an inverse relationship between the soil populations of bacteria and protozoa. Active protozoa were soon demonstrated in soil by Martin and Lewin.[45] After improved methods of counting soil protozoa had been evolved,[9, 10] several investigations of bacterial and protozoan populations in laboratory cultures and under field conditions were initiated. Bacterial populations of one or two species in pre-sterilised soil were often depressed by trophic populations of a single protozoan species.[11, 67] In experiments involving more complex mixtures of microbial species there was often no simple relationship between bacterial and protozoan populations.[70] One notable exception was reported by Cutler, Crump and Sandon.[16] They found that 70% of 339 consecutive daily counts of the common amoeba *Naegleria gruberi* in a well-manured soil (Barnfield at Rothamsted Experimental Station) were significantly inversely correlated with the bacterial population from the same soil samples. It was gradually realised that soil protozoa were not indiscriminate in their choice of food[13, 54–56, 66] and that they were not exclusively bacteriophagous, as assumed by Russell and Hutchinson. Cutler and Crump,[15] for example, commented that the results reported by Cutler, Crump and Sandon[16] '. . . . could only be obtained when the majority of bacteria counted happened to be forms which were readily eaten by the amoebae, and this might not occur in every soil, or on every plating medium. In any case the amoebae cannot be invoked to explain the fluctuations in bacterial numbers that still occur in soils from which all protozoa are absent. . . .' Although such environmental factors as soil temperature and moisture had some effect on the growth and activity of protozoa, it also proved impossible to correlate most of the changes in protozoan populations of field soils with these physical factors.[15] Another objection to the Russell and Hutchinson hypothesis arose from the discovery that the metabolic activity of bacterial cultures were sometimes stimulated rather than inhibited by protozoa. For example, *Azotobacter* spp. can fix more atmospheric nitrogen in the presence of soil protozoa than in their absence.[12, 33, 48] Mixed cultures of soil bacteria and protozoa may also produce more carbon dioxide and ammonia than equivalent monocultures of the bacteria.[14, 46, 47]

In the first two decades after the publication of the Russell and Hutchinson hypothesis considerable progress was made in soil protozoan systematics. Many frequency distribution studies of protozoan species were made in relation to geography, macroclimate and edaphic factors. It proved impossible, however, to correlate protozoan species distribution with particular geographical areas or major soil types or with most macroenvironmental factors. Soil protozoa were found to be ubiquitous; the same species occurred in arctic, temperate and tropical soils. None of the soils examined were completely devoid of protozoa. In arable soils the maximum number of species was usually found at a depth of 10–12 cm below the soil surface and very few species were found in subsoils. According to Sandon,[65] who reviewed most of the frequency distribution studies, except in peaty soils the largest positive correlation occurred between the soil nitrogen content and numbers of species of flagellates, amoebae and ciliates. He suggested that the quantity of available bacterial food was really the most important factor, since '. . . the conditions most favourable to bacterial development are the ones which give rise to the occurrence of the largest numbers of species of these protozoa. Temperature, moisture, reaction and soil texture are of little importance. . . .' The testate amoebae were exceptional because the largest numbers of species occurred in acid peaty soils where other protozoan species were scarce.

The quality of these early studies on soil protozoa can only be fully appreciated if it is remembered how little was then known about the soil microenvironment. The following account attempts to consider soil protozoa as far as possible in terms of the microenvironment. More extensive discussions of the early literature are provided by Koffman[38] and Kopeloff and Coleman.[39] Several reviews of soil protozoology have been published recently.[50, 51, 58, 68, 69, 74]

8.2 Classes and genera of protozoa present in soil

The phylum Protozoa contains four distinct groups or sub-phyla of predominantly microscopic animals: the Sarcomastigophora, Ciliophora, Sporozoa and Cnidospora. The free-living soil protozoa belong to either the first or second sub-phyla; the Sporozoa and Cnidspora are all parasites. The recent textbooks of Dogiel,[20] Grell,[27] Kudo,[41] and Sleigh[72] clearly demonstrate the great diversity and complexity among the Sarcomastigophora and Ciliophora. An outline classification of these two sub-phyla, as proposed by Sleigh,[72] is shown in *Table 8.1* with the orders absent from soils in parentheses.

Table 8.1

CLASSIFICATION OF THE PROTOZOAN SUB-PHYLA SARCOMASTIGOPHORA AND CILIOPHORA ACCORDING TO SLEIGH.[72] THE ORDERS WITHOUT FREE-LIVING SOIL SPECIES ARE SHOWN IN PARENTHESES

Sub-phyla	Classes	Sub-classes	Orders
Sarcomastigophora	Mastigophorea		Cryptomonadida (Haptomonadida) Chrysomonadida Bicoecida Choanoflagellida (Silicoflagellida) (Ebriida) Xanthemonadida Chloromonadida Dinoflagellida Euglenida Phytomonadida Kinetoplastida Metamonadida
	Opalinatea		(Opalinida)
	Sarcodinea	Rhizopodia	Amoebida Testacida
		Granuloreticulosia	(Foraminiferida)
		Mycetozoia	Acrasida
		Actinopodia	Heliozoida (Acantharida) (Radiolarida)
Ciliophora	Ciliatea	Holotricha	Gymnostomatida Trichostomatida (Chonotrichida) (Apostomatida) Astomatida Hymenostomatida (Thigmotrichida)
		Peritrichia	Peritrichida
		Suctoria	Suctorida
		Spirotrichia	Heterotrichida Oligotrichida (Tintinnida) (Entodiniomorphida) (Odontostomatida) Hypotrichida

It can be noted that the soil protozoa are widely distributed among the orders of the Sarcomastigophora and Ciliophora.

An obvious characteristic of all the vegetative cells in the class Mastigophorea is the possession of whip-like locomotory organelles or flagella. This class contains photosynthetic algal flagellates, their colourless relatives and the so-called zooflagellates, which on present evidence do not have obvious affinities with the algal flagellates. Among the phytoflagellates the Cryptomonadida (*Chilomonas, Cryptomonas, Cyathomonas, Rhodomonas*), the Chrysomonadida (*Cephalothamnion, Chrysamoeba, Mallomonas, Monas, Oikomonas, Spongomonas*) and the Euglenida (*Astasia, Distigma, Entosiphon, Euglena, Peranema, Petalomonas, Scytomonas*) are well represented in soil. The common soil zooflagellates are in the Kinetoplastida (*Allantoin, Bodo, Cercobodo, Cercomonas, Helkesimastrix, Heteromita, Pleuromonas, Sainouron, Spiromonas*) and the Choanoflagellida (*Codosiga, Monosiga, Phalansterium, Salpingoeca*).

The Sarcodinea normally possess flexible external protrusions or pseudopodia, which are often used as locomotory organelles. They are placed within the same subphylum as the Mastigophorea, because of some obvious affinities with the latter class. For example, some amoebae possess both flagella and pseudopodia at the same time or at different stages of their life history. Conversely, many flagellates can produce pseudopodia. The sub-class Rhizopodia consists of the naked amoebae (Amoebida) and the testate amoebae (Testacida). The latter are almost entirely enclosed in a shell or test. The common soil genera in the Amoebida are *Acanthamoeba, Amoeba, Hartmanella, Mayorella, Naegleria, Nuclearia*, and *Vahlkampfia*. The common soil genera in the Testacida are *Arcella, Centropyxis, Corythion, Difflugia, Difflugiella, Euglypha, Heleopera, Nebela, Phryganella, Plagiophyxis* and *Trinema*. In the Acrasida (Myxamoebae or slime moulds) and in the Heliozoida the commonest genera in soils are *Dictyostelium* and *Actinophrys*, respectively.

The siliates (Ciliophora) form a natural group and can be clearly distinguished from other protozoa by three major characteristics: the possession of bristle-like locomotory organelles or cilia on their body surface and the associated pellicular organisation, their unique nuclear dimorphism and their method of conjugation. Numerous species are present in the soil and the commonest genera are *Chilodonella, Cinetochilum, Colpoda, Cyclidium, Cyrtolophosis, Dileptus, Enchelys, Gastrostyla, Gonostomum, Halteria, Keronopsis, Oxytricha, Pleurotricha, Prorodon, Saprophilus, Trichopelma, Uroleptus, Urostyla, Uroticha* and *Vorticella*.

8.3 Soil microenvironment

At the microscopical level, soils can be regarded as heterogeneous mixtures of mineral and organic particles with an intricate assortment of pores filled with air or moisture (*Figure 8.1*). Kubiena,[40] the pioneer of micromorphology, obviously grasped the essential heterogeneity of the soil microenvironment when he wrote: '... In the cavities active organisms are to be found, in some only few, in others very many according to the size, climate and food condition of various cavities....' Jongerius[34] classified soil pores into three groups: *micropores* (less than 30 μm diameter), which he thought were important as a moisture reservoir for plants and soil microbes; *mesopores* (diameter between 30 and 100 μm) '... important for intense renewal of air ... serving to transport and distribute water....'; and *macropores* (more than 100 μm diameter), which '... enable air to penetrate the soil rapidly and deeply ...' and allow of '... the rapid withdrawal of large quantities of water....' In many soils a significant proportion of the pores are interconnected by relatively narrow constrictions or 'necks',[5, 71] although some parts of the pore system may be discontinuous.[32]

As the size, shape and arrangement of the constituent particles in a soil determine the number and nature of the soil pores, some differences in dimensions of pore systems between individual soils are to be expected. Salter and Williams[64] studied the relationship between matrix potential and water contents of 20 soils of various textural types. In coarse-textured soils (sands and sand/loamy sands) they found that more than half of the total volume of water retained at saturation was removed at a suction of 0.66 bar, i.e. when all the pores with 'necks' larger than approximately 4 μm diameter assuming a circular outline were drained at 20°C (Griffin).[29] In soils with a finer texture (loamy sands, sandy loams, loams, silt loams and clays) less than half of the volume retained at saturation was removed by this suction. If one assumes that soil pores of less than 4 μm diameter are not penetrated by most soil protozoa, then these results suggest that in the saturated condition a larger proportion of the total volume of the pores in coarse-textured soils are available for protozoan colonisation than in fine-textured soils. Nevertheless, Salter and Williams[64] found that the largest volumes of water, removed from saturated soils at suctions of less than 0.66 bar and consequently the volume available for protozoan colonisation, occurred in silt loams with stable crumb structures. Thus, soil structure affects particularly the frequency of the larger pores. Since

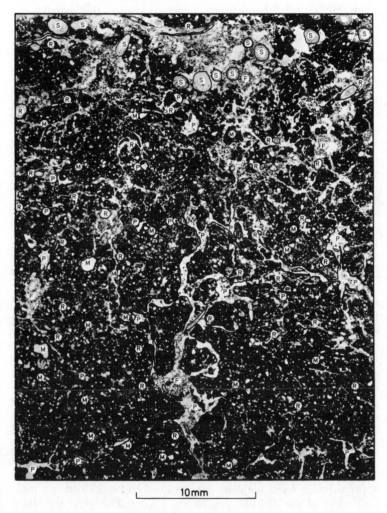

L_____ 10mm _____J

Figure 8.1 Soil thin section showing plant stems (S) and roots (R), mineral particles (M), invertebrate faecal pellets (F) and soil pores (P). Only the larger structures in the photograph are labelled. Not all the numerous small white circular areas represent soil pores. Soil from Dalmeny Park Estate, near Edinburgh; National Grid Reference NT.157776; Darleith series of the Darleith Association; A horizon, 0.2–4 cm; permanent grassland

153

soil structure and the stability of soil aggregates can be altered by many biological, chemical or physical agents, the actual pore network available to soil protozoa must be continuously changing. For example, common agricultural practices, such as cultivation and the addition of bulky organic manures, can significantly alter the structure of many soils.[63] Roots and the macrofauna, especially earthworms, are responsible for many of the macropores.[22, 62] Many soil micro-organisms can produce material that binds soil particles into aggregates.[30] Conversely, microbial decomposition of polysaccharides and humic materials can disperse soil aggregates.[32] Differential weathering of soil materials and the alternation of either freezing and thawing or drying and wetting can all change the size and distribution of soil pores. In addition, illuviation, flocculation or crystallisation of particular soil materials can help to fill the existing pores.[5]

Apart from the relative dimensions of protozoa and soil pores, the most important factor limiting protozoan development is probably the amount of soil solution in the pores. With insufficient moisture, soil protozoa become quiescent and most encyst. Unpublished experiments by the author, using the common soil ciliate *Colpoda steini* and the soil bacterium *Azotobacter chroococcum,* illustrate the importance of moisture in soil pores with dimensions larger than the protozoa *(Figure 8.2).* In each experiment 12 replicate air-dry soil samples (20 g), pre-sterilised by γ-irradiation, were saturated with an *Azotobacter* medium (glucose–mineral salts medium of Brown, Burlingham and Jackson[6]) and inoculated with cultures of *Azotobacter* and *Colpoda.* Subsequently, the soil samples were subjected to one of four suctions (zero, 0.03, 0.10, 0.50 bar) and incubated under controlled conditions to prevent microbial contamination or large environmental changes. The approximate dimensions of the *Colpoda* and *Azobacter* cells used in these experiments were $30 \times 19 \times 10\ \mu m$ and $3.8 \times 2.7 \times 2.7\ \mu m$ respectively. The microbial populations in the soil samples were estimated by dilution methods.[18, 19]

At a suction of 0.50 bar, when all the pore 'necks' of more than approximately 6 μm diameter are assumed to be drained, no multiplication of the *Colpoda* population occurred. Indeed the slight decrease in the *Colpoda* population after inoculation suggests that some ciliates were destroyed at this suction and did not encyst. In the other experiments with lower suctions the ciliates were able to multiply and the duration of multiplication was inversely related to the suction applied. The largest final populations of *Colpoda* were attained in saturated soils. The maximum diameter of the pore

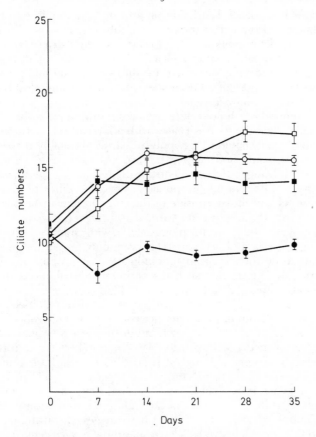

Figure 8.2 Colpoda steini *populations inoculated into soil samples, pre-sterilised by γ-irradiation, and subjected to one of four suctions, in bars:* ●——● *0.50;* ■——■ *0.10;* ○——○ *0.03;* □——□ *zero. The ciliate populations per gram oven-dry soil are expressed in logarithms to the base 2. 95% fiducial limits for each population estimate are indicated*

'necks' not drained by suctions of either 0.03 or 0.10 bar were 94 μm and 30 μm, respectively, at the temperature used for these incubations. The *Azotobacter* populations multiplied successfully in all soil samples within the range of suctions tested and probably some of this growth occurred in the smaller pores isolated from the *Colpoda* population. Soil pores which cannot be drained except through channels smaller than the minimal diameters listed above will remain full of soil solution.

Another important restriction on protozoan multiplication in the soil microenvironment is the discontinuous distribution of organic matter and the microflora.[5, 36, 49] Existing evidence also suggests that most of the organic matter and microflora are associated with soil aggregates and not dispersed in the soil solution.[2, 25, 44] It is impossible at present to estimate what proportion of the food associated with soil aggregates is available to protozoa. There is no information about how readily protozoa can remove sorbed food, especially when the food is coated with a layer of clay particles[44] or alien microbial material, e.g. polysaccharide. Probably the central regions of many soil aggregates are inaccessible to protozoa. Hattori[32] has suggested, mainly from the results of experiments on the effects of dispersants on soil aggregates, that different microbes colonise the outside and inside of these aggregates. In his view the outside of the aggregates are colonised chiefly by fungi, actinomycetes and spore-forming bacteria, with Gram-negative non-sporing bacteria and actinomycetes predominating in the interior. Leaving aside the important question of what proportion of soil protozoa are saprozoic,[3] such discontinuous distributions suggest that soils often contain significant food reserves, which are unavailable to protozoa at any given moment. Consequently, protozoa, in common with other heterotrophic soil microbes,[26] may sometimes exist in soil under conditions of nutrient limitation. The small size of many soil protozoa[38, 40, 65] may reflect the nutritional deficiencies of soils, although it is possible that the soil forms are distinct genotypes from the same species found in other environments. It is established, however, that the starvation of laboratory cultures of protozoa can induce the formation of dwarf populations.[15, 31] Stout and Heal[74] have suggested, on the basis of Adolph's experiments with *Colpoda* sp. at low oxygen tensions, that small protozoa in the soil may result from reduced oxygen tensions as well as from reduced food supply. It is also known that temperature can influence the size of protozoa.[35, 75]

Aeration, temperature and pH are probably the most important physicochemical factors apart from moisture affecting soil protozoa. In general, soil atmospheres contain less oxygen and more carbon dioxide than adjacent above-ground atmospheres. Nitrogen is present in almost the same concentrations in both atmospheres.[43] Under conditions of impeded gas exchange an accumulation of gaseous metabolites, e.g. ethylene or hydrogen sulphide, may occur.[73] Soil temperature fluctuations except near the surface tend to be smaller than above-ground temperatures.[76] McLaren and

Skujins[43] have rightly stressed that local variations and complex interactions of these physical factors occur in the soil microenvironment. A decrease in soil moisture not only increases soil aeration[8] but may also cause some shrinkage of the soil pores and an increase in the concentration of the soil solution. The solubility of carbon dioxide or oxygen in the soil solution is affected by temperature fluctuations and salt concentrations. Plant roots or soil animals can create local differences in pH and in partial pressures of oxygen.[28, 53] Intense local microbial activity near plant roots or organic matter can induce local gradients in temperature, pH or oxygen tension.[7]

8.4 Physiological and morphological characteristics of soil protozoa

Protozoan existence in the complex soil microenvironment necessitates a wide degree of tolerance to environmental conditions. Noland and Gojdics[52] thought a facile capacity for rapid encystment and excystment was a vital characteristic of soil protozoa. Numerous experiments have shown the resistance of protozoan cysts to extremes of desiccation, temperature, pH and the gaseous environment. Many active soil protozoa also show a wide tolerance to these environmental factors.[37, 52, 57, 74] Testate amoebae are further protected from their environment by their external shells or tests, and in some soil species the orifice of their test is small.[4] Sandon[65] suggested that the inability to encyst might explain the absence in soils of such common freshwater genera as *Paramecium,* although it does not explain the absence or rarity of other common freshwater genera in soils.

In common with other soil fauna, protozoan species are characterised by their small size.[3, 42] Even within a single protozoan genus, the edaphic species are often smaller than the common freshwater or marine species.[74] There is usually a lack of external spines or projections[65] and many soil protozoa are flattened to some degree.[4, 74] The presence of similar morphological features among the ciliates inhabiting the fine pores of littoral marine sands[21, 24] suggests that they are advantageous in the restricted porous environments of both marine sands and soils. Dragesco[21] also noted that many ciliates in fine marine sands were characterised by their ciliation being restricted to one side of the body and by positive thigmotaxis. Such ciliates (Hypotrichida) are well represented in soils. Many soil flagellates and amoebae exhibit positive thigmotaxis on microscope slides and a similar response in soils may be an important characteristic for survival, especially if these protozoa can attach themselves to edible substrates. It is possible that the long trailing flagella of many soil flagellates[65] may have an important tactile function. Soil ciliates are potentially more mobile in the soil than flagellates or amoebae. In soil suspensions, however, many ciliates can be observed around deposits of organic matter or concentrations of the microflora and are not free-swimming. It is probable that such a situation exists in undisturbed soils. Stout and Heal[74] observed that, apart from a single species of *Vorticella,* no common soil ciliate is attached to the substrate. Fauré-Fremiet[23] observed a similar absence of sessile ciliates in marine littoral sands.

Sandon[65] concluded from the results of a world survey that the majority of soil protozoan species were also common in several freshwater and marine habitats rich in organic matter and that only 21 species were endemic to soils. More recent investigations have extended the species list but this general conclusion still stands. Such a community is likely to provide some colonists for a range of ecological conditions and would appear to be well suited to life in soils. The absence, however, of several common freshwater protozoa from soils supports Sandon's view that soil protozoa are not an accidental collection of less specialised types. Unfortunately, the distinctive physiological requirements of soil protozoa are still poorly comprehended.

8.5 The significance of protozoa in the soil community

With the advantage of our present knowledge of the soil microenvironment, it is not surprising that early attempts to correlate the size of protozoan populations with macroclimatic conditions or total bacterial populations were largely unsuccessful. A more promising approach now appears to be to consider the fluctuations in active populations of soil protozoa in terms of the volume of moisture present in the soil pores. Soil moisture is an essential requirement for active protozoa and its presence or absence has a major influence on the microenvironment. Approximate measurements of the total volume of the saturated pores available for protozoan multiplication in non-shrinking soils could be obtained by monitoring changes in soil water tension and from a knowledge of the moisture characteristics of undisturbed soil cores. The practical difficulties of these measurements will be greatest near the surface and where the soil moisture gradients are large. Such an approach could be a basis for comparing the suitability of the microenvironments of different soils and different soil horizons for protozoan growth, if the minimal pore size that can be penetrated by different species of protozoa is known. Microscopical observations of ciliates moving among soil particles on microscope slides suggest that many of these protozoa can squeeze through some apertures smaller than their own normal dimensions. Naked amoebae are clearly very flexible organisms, and Bonnet[4] has observed that pseudopodia of testate amoebae can penetrate thin films of water, which are too small to contain their tests. Further microscopical observations should provide the necessary information about the minimal size of soil pores accessible to soil protozoa.

It is premature to ascribe any more specific role to soil protozoa than an involvement in organic matter decomposition and in the predation of the microflora until there is much more knowledge of how individual protozoan species behave in the microenvironment. Additional autecological studies combined with modern methods of estimating soil populations would seem to be a promising trend for future protozoological research.

Acknowledgements

The author wishes to thank Mr. L. Robertson, Department of Soil Survey of the Macaulay Institute for *Figure 8.1*. Several other colleagues and former colleagues contributed constructive criticisms to the text.

REFERENCES

1. ADOLPH, E. F. The regulation of adult body size in the protozoan Colpoda, *Journal of Experimental Zoology*, **53**, 269–311 (1929)
2. BACON, J. S. D. The chemical environment of bacteria in soil, in *The Ecology of Soil Bacteria* (edited by T. R. G. Gray and D. Parkinson), 25–43. Liverpool University Press (1968)
3. BECK, T. *Mikrobiologie des Bodens*. Bayerischer Landwirtschafts-Verlag, München (1968)
4. BONNET, L. Le peuplement thécamoebien de sols, *La Revue d'Écologie et de Biologie du Sol*, **1**, 123–408 (1964)
5. BREWER, R. *Fabric and Mineral Analysis of Soils*. Wiley, New York (1964)
6. BROWN, M. E., BURLINGHAM, S. K. and JACKSON, R. M. Studies on *Azotobacter* species in soil. I. Comparison of media and techniques for counting Azotobacter in soil, *Plant and Soil*, **17**, 309–319 (1962)

159

7. CLARK, F. E., JACKSON, R. D. and GARDNER, H. R. Measurement of microbial thermogenesis in soil, *Proceedings of the Soil Science Society of America*, **26**, 155–160 (1962)

8. CURRIE, J. A. Gaseous diffusion in porous media. Part 3. Wet granular materials, *British Journal of Applied Physics*, **12**, 275–281 (1961)

9. CUTLER, D. W. Observations on soil protozoa, *Journal of Agricultural Science* (Cambridge), **9**, 430–444 (1919)

10. CUTLER, D. W. A method for estimating the number of active protozoa in the soil, *Journal of Agricultural Science* (Cambridge), **10**, 135–143 (1920)

11. CUTLER, D. W. The action of protozoa on bacteria when inoculated into sterile soil, *Annals of Applied Biology*, **10**, 137–141 (1923)

12. CUTLER, D. W. and BAL, D. V. Influence of protozoa on the process of nitrogen fixation by *Azotobacter chroococcum*, *Annals of Applied Biology*, **13**, 516–534 (1962)

13. CUTLER, D. W. and CRUMP, L. M. The qualitative and quantitative effects of food on the growth of a soil amoeba (*Hartmanella hyalina*), *British Journal of Experimental Biology*, **5**, 155–165 (1927)

14. CUTLER, D. W. and CRUMP, L. M. Carbon dioxide production in the sands and soils in the presence and absence of amoebae, *Annals of Applied Biology*, **16**, 472–482 (1929)

15. CUTLER, D. W. and CRUMP, L. M. *Problems in Soil Microbiology*. Longmans Green, London (1935)

16. CUTLER, D. W., CRUMP, L. M. and SANDON, H. A quantitative investigation of the bacterial and protozoan population of the soil, with an account of the protozoan fauna, *Philosophical Transactions of the Royal Society of London*, Series B, **211**, 317–350 (1922)

17. DARBISHIRE, F. V. and RUSSELL, E. J. Oxidation in soils and its relation to productiveness. Part II. The influence of partial sterilisation, *Journal of Agricultural Science (Cambridge)*, **2**, 305–326 (1908)

18. DARBYSHIRE, J. F. Nitrogen fixation by *Azotobacter chroococcum* in the presence of *Colpoda steini*. I. The influence of temperature, *Soil Biology and Biochemistry*, **4**, 359–369 (1972)

19. DARBYSHIRE, J. F., WHEATLEY, R. E., GREAVES, M. P. and INKSON, R. H. E. A rapid micromethod for counting bacteria and protozoa in soil, *La Revue d'Écologie et de Biologie du Sol*, **11**, 465–475 (1974)

20. DOGIEL, V. A. *General Protozoology* (revised by J. I. Poljansky and E. M. Chejsin). Oxford University Press (1965)

21. DRAGESCO, J. Ciliés mésopsaminques littoraux. Systématique, morphologie, écologie, *Des Travaux de la Station Biologique de Roscoff*, **12**, 11–356 (1960)

22. EDWARDS, C. A. and LOFTY, J. R. *Biology of Earthworms*. Chapman and Hall, London (1972)

23. FAURÉ-FREMIET, E. Écologie des ciliés psammophiles littoraux, *Bulletin Biologique de la France et de la Belgique*, **84**, 35–75 (1950)

24. FAURÉ-FREMIET, E. The marine sand-dwelling ciliates of Cape Cod, *Biological Bulletin of the Marine Biology Laboratory, Woods Hole, Massachusetts*, **100**, 59–70 (1951)

25. GRAY, T. R. G., BAXBY, P., HILL, I. R. and GOODFELLOW, M. Direct observation of bacteria in soil, in *The Ecology of Soil Bacteria* (edited by T. R. G. Gray and D. Parkinson), 171–192. Liverpool University Press (1968)

26. GRAY, T. R. G. and WILLIAMS, S. T. Microbial productivity in soil, in *Microbes and Biological Productivity* (edited by D. E. Hughes and A. H. Rose), 255–286. Cambridge University Press (1971)

27. GRELL, K. G. *Protozoology*. Springer-Verlag, Berlin (1973)

28. GREENWOOD, D. J. Studies on the distribution of oxygen around the roots of mustard seedlings (*Sinapis alba* L.), *New Phytologist,* **70**, 97–101 (1971)

29. GRIFFIN, D. M. Soil moisture and the ecology of soil fungi, *Biological Reviews,* **38**, 141–166 (1963)

30. GRIFFITHS, E. Micro-organisms and soil structure, *Biological Reviews,* **40**, 129–142 (1965)

31. HARDING, J. P. Quantitative studies on the ciliate *Glaucoma* I. The regulation of the size and the fission rate by the bacterial food supply, *Journal of Experimental Biology,* **14**, 422–430 (1937)

32. HATTORI, T. *Microbial Life in Soil: An Introduction.* Marcel Dekker, New York (1973)

33. HIRAI, K. and HINO, I. Influence of soil protozoa on nitrogen fixation by *Azotobacter, Proceedings of the First International Congress of Soil Science, Washington,* Vol. 3, 160–165 (1928)

34. JONGERIUS, A. *Morphologic Investigations of Soil Structure.* Bodemkundige Studies No. 2, Mededelingen van die Stickting voor Bodemkartering, Wageningen (1957)

35. JOHNSON, B. F. Influence of temperature on the respiration and metabolic effectiveness of *Chilomonas, Experimental Cell Research,* **28**, 419–423 (1962)

36. JONES, D. and GRIFFITHS, E. The use of thin soil sections for the study of soil micro-organisms, *Plant and Soil,* **20**, 232–240 (1964)

37. KITCHING, J. A. Some factors in the life of free-living protozoa, in *Microbial Ecology* (edited by R. E. O. Williams and C. C. Spicer), 259–286. Cambridge University Press (1957)

38. KOFFMAN, M. Die Mikrofauna des Bodens, ihr Verhältnis zu anderen Mikroorganismen und ihre Rolle bei den mikrobiologischen Vorgängen im Boden, *Archiv für Mikrobiologie,* **5**, 246–302 (1934)

39. KOPELOFF, N. and COLEMAN, D. A. A review of investigations in soil protozoa and soil sterilization, *Soil Science,* **3**, 197–269 (1917)

40. KUBIENA, W. L. *Micropedology.* Collegiate Press, Ames (1938)

41. KUDO, P. R. *Protozoology,* 5th edition. Thomas, Springfield (1966)

42. KÜHNELT, W. An introduction to the study of soil animals, in *Soil Zoology* (edited by D. K. M. Kevan), 3–22. Butterworths, London (1955)

43. McLAREN, A. D. and SKUJINS, J. The physical environment of micro-organisms in soil, in *The Ecology of Soil Bacteria* (edited by T. R. G. Gray and D. Parkinson), 3–24. Liverpool University Press (1968)

44. MARSHALL, K. C. Sorptive interactions between soil particles and microorganisms, in *Soil Biochemistry,* Volume 2 (edited by A. D. McLaren and J. Skujins), 409–445. Marcel Dekker, New York (1971)

45. MARTIN, C. H. and LEWIN, K. R. Notes on some methods for the examination of soil protozoa, *Journal of Agricultural Science,* (Cambridge), **7**, 106–119 (1915)

46. MEIKLEJOHN, J. The relation between the numbers of a soil bacterium and the ammonia produced by it in peptone solutions; with some reference to the effect on this process of the presence of amoebae, *Annals of Applied Biology,* **17**, 614–637 (1930)

47. MEIKLEJOHN, J. The effect of *Colpidium* on ammonia production by soil bacteria, *Annals of Applied Biology,* **19**, 584–608 (1932)

48. NASIR, S. M. Some preliminary investigations on the relationship of protozoa to soil fertility with special reference to nitrogen fixation, *Annals of Applied Biology,* **10**, 122–133 (1923)

49. NICHOLAS, D. P., PARKINSON, D. and BURGES, N. A. Studies of fungi in a podzol. II. Application of the soil-sectioning technique to the study of amounts of fungal mycelium in the soil, *Journal of Soil Science,* **16**, 258–269 (1965)

50. NICOLJUK, V. F. *Soil Protozoa and their Role in Cultivated Soils of Uzbekistan* [in Russian]. Izdatel'stvo Akademii Nauk Uzbekskoi SSR, Tashkent (156)

51. NICOLJUK, V. F. and GELTZER, J. G. *Soil Protozoa of the USSR* [in Russian]. Akademii Nauk Uzbekskoi SSR, Tashkent (1972)

52. NOLAND, L. E. and GOJDICS, M. Ecology of free-living protozoa, in *Research in Protozoology,* Volume 2 (edited by T. Chen), 215–266. Pergamon Press, London (1967)

53. NYE, P. H. Processes in the root environment, *Journal of Soil Science,* **19,** 205–215 (1968)

54. OEHLER, R. Amoebenzucht auf reinem Boden, *Archiv. für Protistenkunde,* **37,** 175–190 (1916)

55. OEHLER, R. Weitere Mitteilungen über gereingte Amoeben und Ciliatenzucht, *Archiv für Protistenkunde,* **49,** 112–134 (1924)

56. OEHLER, R. Gereinigte Zucht von freilebenden Amoeben, Flagellaten und Ciliaten, *Archiv für Protistenkunde,* **49,** 287–296 (1924)

57. POLJANSKY, G. I. The problem of physiological adaptation with regard to the forms of variability in free living protozoa (some results and perspectives), in *Progress in Protozoology,* 40–55. Université de Clermont, UER-Sciences Exactes et Naturelles, Clermont–Ferrand (1973)

58. PUSARD, M. Les protozoaires du sol, *Annales des Épiphyties,* **18,** 335–360 (1967)

59. RUSSELL, E. J. Oxidation in soils, and its connexion with fertility, *Journal of Agricultural Science* (Cambridge), **1,** 261–279 (1905)

60. RUSSELL, E. J. and HUTCHINSON, E. B. The effect of partial sterilisation of soil on the production of plant food, *Journal of Agricultural Science* (Cambridge), **3,** 111–144 (1909)

61. RUSSELL, E. J. and HUTCHINSON, E. B. The effect of partial sterilisation of soil on the production of plant food. II. The limitation of bacterial numbers in normal soils and its consequences, *Journal of Agricultural Science* (Cambridge), **5,** 152–221 (1913)

62. RUSSELL, E. W. *Soil Conditions and Plant Growth,* 9th edition. Longmans Green, London (1961)

63. SALTER, P. J. and WILLIAMS, J. B. The effect of farmyard manure on the moisture characteristic of a sandy loam soil, *Journal of Soil Science,* **14,** 73–81 (1963)

64. SALTER, P. J. and WILLIAMS, J. B. The influence of texture on the moisture characteristics of soils. II. Available-water capacity and moisture release characteristics, *Journal of Soil Science,* **16,** 310–317 (1965)

65. SANDON, H. *The Composition and Distribution of the Protozoan Fauna of the Soil.* Oliver and Boyd, Edinburgh (1927)

66. SANDON, H. *The Food of Protozoa.* Misr-Sokkar Press, Cairo (1932)

67. SINGH, B. N. Selectivity in bacterial food by soil amoebae in pure mixed culture and in sterilised soil, *Annals of Applied Biology,* **28,** 52–64 (1941)

68. SINGH, B. N. Inter-relationship between micropredators and bacteria in soil, *Proceedings 47th Indian Science Congress,* Part 2, 1–14 (1960)

69. SINGH, B. N. Recent advances in soil protozoology, *Proceedings of the First Summer School of Zoology at Simla 1961,* 79–103 (1963)

70. SINGH, B. N. and CRUMP, L. M. The effect of partial sterilization by steam and formalin on the numbers of amoebae in field soil, *Journal of General Microbiology,* **8,** 421–426 (1953)

71. SLATYER, R. O. *Plant–Water Relationships.* Academic Press, New York (1967)

72. SLEIGH, M. *The Biology of Protozoa.* Edward Arnold, London (1973)

73. SMITH, K. A. and RESTALL, S. W. F. The occurrence of ethylene in anaerobic soil, *Journal of Soil Science,* **22,** 430–443 (1971)

74. STOUT, J. D. and HEAL, O. W. Protozoa, in *Soil Biology* (edited by N. A. Burges and F. Raw), 149–195. Academic Press, London (1967)

75. THORMAR, H. Effect of temperature on the reproduction rate of *Tetrahymena pyriformis, Experimental Cell Research,* **28,** 269–279 (1962)

76. WEST, E. S. A study of the annual soil temperature wave, *Australian Journal of Scientific Research,* Series A, **5,** 303–314 (1952)

9 Soil Fungi

R. M. JACKSON

Early work on soil fungi was mainly concerned with counting and cataloguing those fungi that could be isolated from soil. The reasons for this are clear: techniques already existed which, with minor modifications, made such studies relatively simple and the work was rewarding because it gave evidence of the presence of a variety of fungi in soil, apparently in great numbers. Quite early during this phase, doubts arose as to whether fungi were really growing in soil or were there accidentally as contaminants from other substrates. Waksman[59] attempted to resolve this problem by isolating fungi that grew from large soil inocula in 24 h, arguing that such fungi must originate from hyphae rather than spores. It is doubtful whether this was correct,[62] but several more sophisticated techniques have amply confirmed that fungi do grow in soil.

During the last 25 years interest has turned largely towards understanding in what form and quantity fungi are present in soil, how active they are and in what kinds of transformations they are important. Most studies have been synecological and there have been few autecological studies of non-pathogenic soil fungi. An example of the latter is the work of Danielson and Davey[15] on *Trichoderma* spp. Progress has been limited by the slow development of new techniques, their apt application and the interpretation of the data obtained. That good techniques may not be quickly superseded by better ones is well illustrated by the agar film technique described by Jones and Mollison[28] in 1948. This simple but elegant technique yields a partial picture of the kinds and quantities of micro-organisms present in soil. Since its introduction this method, or modifications of it (see, for example, reference 55), have been and are still used by numerous workers, giving results not readily obtained by any other method. Their chief limitation, as with similar methods depending on direct microscopy, is in failing to discriminate between organisms that were alive when the sample was taken and those that were dead. None of the staining methods tested

for this purpose has proved entirely satisfactory; indeed, interpretation of results from staining depends on assumptions that are
difficult or impossible to test.

A better criterion of viability than uptake of certain stains is
metabolic activity. The uptake of radioisotopes such as ^{14}C and [3]H
from their environment followed by autoradiography has been used
to distinguish living from dead algal and bacterial cells in sea-
water.[7-9] This technique has been adapted to detect metabolically
active micro-organisms in leaf litter, by exposing the litter for a short
period to uniformly labelled [^{14}C]glucose and subsequently making
autoradiographs.[57] It was assumed that hyphae taking up the label,
i.e. assimilating glucose, were living, while those giving no
autoradiogram must be either dead or dormant. There must be
certain reservations about this assumption, since little is known
about the uptake of glucose or other organic compounds from soil
by important groups of fungi, such as many mycorrhizal fungi.
That some otherwise active fungi may not be able to take up glucose
is a real possibility; thus, Hacskaylo[21] concludes that in nature
sheathing mycorrhizal fungi depend primarily upon roots of their
hosts for carbohydrates, including glucose.

The value of the Jones and Mollison method would be greater if it
were combined with an autoradiographic technique to differentiate
metabolically active organisms. Another potential development is
fluorescent antibody (FA) staining of Jones and Mollison films to
identify particular species. The technique of FA staining has recently
been successfully applied to determining bacterial growth rates in
soil[5] and doubtless will be increasingly used in soil microbiology.
While dilution plates grossly underestimate fungi present in soil as
hyphae, Jones and Mollison's method is likely to overestimate them.
Warcup's painstaking study[60] of the origins of fungi on soil dilution
plates showed that many colonies originate from chlamydospores
embedded in fragments of organic matter. A high proportion of
such fragments might be expected to be discarded in making Jones
and Mollison films, and even if present, spores in these fragments
would be very difficult to see.

Another criterion of life is the ability of a fungal hypha or
propagule to initiate new hyphal growth. Jones and Mollison films
can be prepared in dilute nutrient agar and incubated for a short
period before being fixed and stained. New growth is distinguished
when these incubated films are examined. The viability and
sometimes identity of hyphae can also be determined if they are
extracted from soil and plated on nutrient media. Values of up to

75% viability of such hyphae have been recorded, but are usually much lower.[27, 61]

Biomass determinations based on the estimation of various chemical compounds have been described in recent years. These include ATP,[2, 24] muramic acid[37] and chitin.[54] Only the latter method is specifically relevant to fungi. Its accuracy depends on calibration with pure cultures and on knowledge of the stability of chitin following death of hyphae. The presence of significant amounts of arthropod chitin would invalidate this method. A novel approach to estimating soil biomass involves measurement of the size of the flush of decomposition that follows partial sterilisation of soil. This method, fully discussed by Powlson in another chapter, cannot, however, differentiate between different parts of the biomass.

Before discussion of the study of the forms and condition of fungi in soil is continued it will be helpful to classify them arbitrarily as follows.

I. Dead: Little is known about how long dead fungal structures may survive in a recognisable form in soil. Dead hyphae extracted from a pasture soil and then suspended in nylon bags in nutrient solutions with an unsterile soil inoculum were recovered intact and apparently little changed after from 3 to 6 months.[27] This suggests a degree of persistence of fungal cell walls that makes it certain that many hyphae seen in such a soil must be dead.

II. Live: (a) Metabolically active and growing. (b) Metabolically active, but not growing. (c) Metabolically inactive, but capable of immediate growth when removed from soil. (d) Metabolically inactive and incapable of growth when removed from soil except following special treatment.

The first of these four categories comprises hyphae and developing reproductive structures. The second will also contain hyphae and probably young and short-lived spores. Any structures in these two categories may be isolated and grown directly or 'grown on' following extraction from soil, provided certain conditions are met. The first is that the method used has not damaged the propagules so as to prevent subsequent growth. The second is that the cultural conditions provided should allow the full potential for growth to be expressed. The third is that the structures are of fungi that are intrinsically culturable, i.e. neither true obligate parasites nor obligate symbionts. The major group of live but non-culturable fungi in the soil is the mycorrhizal fungi. The presence of vesicular–arbuscular mycorrhizal fungi among other hyphae extracted from soil has been

167

observed[27,61] and it is likely that they constitute a significant proportion of all hyphae in a soil. While some sheathing mycorrhizal fungi can be isolated rather easily, at least from mycorrhizal roots, others have proved difficult or impossible to isolate. Zak and Marx[64] reported that isolation from mycorrhizal roots of slash pines was successful in only 40% of attempts. Isolation from sporophore tissue of some mycorrhizal Hymenomycetes is equally difficult. Successful 'growing on' of single hyphae of any of these fungi could not be expected. It is certainly true that, once in culture, some Hymenomycetes can be sub-cultured successfully only by using large inocula.

Categories (c) and (d) comprise mainly spores, but (c) may also include sclerotia and viable, but inactive, mycelium. The terms metabolically active and metabolically inactive have meaning only in a relative sense and represent two halves of a spectrum ranging from dead to extreme activity. Live, but inactive, cells differ from dead cells in possessing a real, but low, respiration rate, intact osmotic barriers and an ATP pool.[43] The condition of structures in category II(c) is one of exogenous dormancy and that of structures in II(d) constitutional dormancy.[53]

Exogenous dormancy or dormancy imposed by the environment is now known to be the usual condition of most fungal propagules in soil. Pioneer studies of fungal spores in soil were made by Dobbs and Hinson,[16] who, in 1953, described what they termed 'widespread fungistasis in soil'. This report drew attention to a phenomenon, previously observed by some workers and doubtless surmised by others. Although there has been a considerable division of opinion as to the causes of soil fungistasis, few soil microbiologists have doubted its existence or its survival value, particularly to fungi that normally inhabit the rhizosphere or are root parasites. Two apparently conflicting hypotheses have been proposed to explain soil fungistasis. The first is that failure of spores to germinate in soil, except when their basic requirements are not satisfied, results from the presence of inhibitory substances of microbial origin. The main weakness of this hypothesis has been the difficulty of demonstrating the presence of such inhibitors in solutions expressed from soil. When such substances have been detected, they have tended to be fugitive. A different hypothesis was proposed by Lockwood[35] and his associates at Michigan. According to their 'nutrient depletion' theory, the soil behaves as a nutrient sink: spores introduced into it rapidly lose nutrients to the environment until they are no longer capable of germination, unless fresh supplies of nutrients become available. Undoubtedly, nutrients are lost from spores in soil or exposed to diffusion gradients in model systems, but the relevance of this phenomenon to soil fungistasis is not clear. There are difficulties in accepting 'nutrient depletion' as a main cause of fungistasis; one piece of experimental work illustrates this. Knight[30] used macroconidia of *Cochliobolus sativus* labelled with ^{14}C to study loss of carbon compounds and its relation to germinability. He placed conidia on the surface of dialysis tubing filled with water or soil suspension to simulate the 'nutrient depletion' occurring in soil. Although there was heavy loss of labelled carbon compounds from the conidia, their germination when transferred to soil was better than that of untreated conidia. Removal of self-inhibitors, in addition to nutrients, may have occurred on the dialysis tubing.

Recently, because of the advent of gas chromatography, interest has been focused on volatile inhibitors of fungi in soil. There is now good evidence that volatile compounds produced by soil organisms can cause fungistasis.[25, 26] Smith[51] detected ethylene production in a range of soils and demonstrated good correlation between the ethylene produced and the level of fungistasis. It remains to be seen whether ethylene alone can account for all normal soil fungistasis.

Thus, the weight of evidence now seems to be favouring an important role for inhibitors of microbial origin in the exogenous dormancy of spores in soil. However, the nutrient status of both the fungal propagule and the environment clearly will strongly interact with such inhibitors. If such nutrient interactions are ignored in work on fungistasis, the results obtained may be misleading. Work in the author's laboratory has shown that the substrate on which a spore is produced affects both its nutrient status and its subsequent response to fungistasis. A practical difficulty in work on fungal spore germination is in harvesting and handling spores without causing nutrient changes. If spores are harvested from the surface of an agar culture in an aqueous medium, then there is a danger that they will be contaminated by unused nutrients from the substrate. Washing the spores is clearly no remedy if this removes nutrients from them. With dry-spored fungi these problems can be avoided by harvesting the spores dry with a vacuum collector, such as that used by Knight,[30] and depositing them directly on the surface where their behaviour is to be studied.

True constitutional or endogenous dormancy has been demonstrated for rather few soil fungi. The classical study was that of Blackwell[4] on *Phytophthora cactorum*. Oospores of this fungus pass through three distinct stages of development before germination is possible. Depending upon conditions, this may take a total of 3–10 months. Blackwell discovered that individual spores differed widely in their period of dormancy and in their response to changes in environmental conditions. Marked heterogeneity among a dormant spore population would help to ensure that some individuals would respond to stimulation while others would remain ungerminated to respond to later stimulation. It is believed that constitutional dormancy must have great survival value for root pathogens, particularly if they have a limited host range and long periods occur when no suitable host is available. In fact, little is known of the occurrence and behaviour in soil of the oospores of some important pathogens. It is not known, for example, whether oospores have a significant role in the survival of the destructive parasite *Phytophthora cinnamomi,* although the fungus forms oospores in soil.[38, 47]

This discussion of the condition of fungi in soil started with the problems inherent in measuring fungal tissue (biomass) and its activity. If the aim is to measure activity over a period of time, different approaches from those already discussed are available. The total CO_2 output or O_2 uptake of a soil can be measured with a suitable respirometer but the value obtained will represent the summation of respiration of plant roots, soil animals, bacteria, fungi and other micro-organisms. The contribution of fungi cannot be distinguished from that of other organisms, so soil respiration studies are of value in assessing fungal activity only in soils where such activity predominates.

A parameter of activity, of at least some fungi, is hyphal extension through soil. An inert surface or complex of surfaces is buried in soil for a given time and then removed for microscopic examination. The Rossi–Cholodny slide is a simple version of such a technique. Nylon mesh has been used in a comparable way by Waid and Woodman[58] and more recently by Nagel-de Boois.[40] The nylon monofilaments of the mesh used provide a continuous reticulum over which hyphae may grow. Because of their geometry and surface properties, the filaments may well encourage more hyphal growth of some species than would otherwise occur. Also, it is very likely that the activity recorded by using buried nylon mesh will not be representative of all hyphal types. Gray and Williams[17] classified patterns of hyphal growth in soil. Fungi in two of the categories they recognised are most likely to colonise buried nylon mesh. One of these categories contains fungi that have mycelial 'strands' or rhizomorphs, by means of which they grow through the soil from substrate to substrate.

Individual hyphae frequently radiate from rhizomorphs and 'strands'. Fungi with this mode of growth are usually abundant in woodland and forest soils, the majority being litter-decomposing and mycorrhizal Hymenomycetes and Gasteromycetes. Fungi with a diffuse spreading hyphal pattern are also likely to be frequent colonisers of buried nylon mesh. These fungi are not necessarily attached to nutrient substrates, but may obtain nutrients from the soil solution. Probably rarely represented on nylon mesh will be fungi with restricted hyphal growth. Organisms of this kind produce dense hyphal growth on and within small pieces of substrate with little or no extension into the soil pores. If a more or less constant proportion of the growth of one hyphal type in a given volume of soil occurs on the mesh, then useful information can be gained of differences in growth activity of that fungus over different periods. Antagonism or hyphal interference between species on the mesh could give spurious results.

Developing interest in fungal behaviour in soil has led to a more critical approach to the study of the influence of the physical environment on fungal activity. In much of the older work on soil fungi the influence of water was either ignored or not properly understood. It is largely due to Griffin's thoughtful experimental work and critical reviews[18-20] that we now have a much clearer understanding of how water affects soil organisms. Cook and Papendick[11-13] reviewed the influence of soil water potential on root disease organisms and also experimented in this field.

The most easily measured parameter of water in a soil is the water content. This, the total water present, is usually expressed as grams of water per gram of soil or as the volume of water per cubic centimetre of soil. Another form of expression frequently used is as a percentage of saturation or water-holding capacity. Knowledge of the total water present in a soil has some value if work is confined to one soil and comparisons are made in that soil under differing conditions. However, the most significant property of soil water for fungi—as for other organisms in soil, including plant roots—is the force holding the water in a soil at any one time, which must be overcome before water can be withdrawn from the soil. The total forces to which soil water is subjected is usually termed total potential or water potential and expressed as a pF value ($\log_{10}$ of negative pressure in centimetres of water) or in bars. Unfortunately, the accurate measurement of soil water potential, except over a limited range, is difficult and requires the use of a pressure membrane apparatus. Once a moisture characteristic curve relating water content to water potential has been constructed, desired water potentials can be easily produced and maintained. A practical difficulty in work on soil moisture is encountered in attempting to moisten a soil evenly to a given potential without excessive breakdown of the soil structure. A way to overcome this difficulty is to mix a calculated quantity of ground ice with the soil at below freezing point (H. Schüepp, personal communication; reference 45, 46). Much experimental evidence shows that fungi are active at water potentials far lower than those that prevent bacterial growth. Griffin[20] concluded that the lower limit for bacterial growth in soil imposed directly by water potential is probably at about −80 bar. Chen and Griffin[10] obtained good colonisation by fungi of short lengths of hair placed on the soil surface at water potentials down to −306 bar, and some colonisation at −395 bar. However, colonisation was much slower at low water potentials than at higher ones. Below −145 bar potential almost the only active fungi were species of *Aspergillus* and *Penicillium*. Significantly, Chen and Griffin detected no difference in the prevalence of species able to grow at −300 bar between soils from an English pasture, an Australian rain forest or a desert.

Interactions between organisms can lead to apparent anomalies in the influence of water potential on behaviour. Thus, *Fusarium roseum* f. sp. *cerealis* 'culmorum' grows best in sterile soil at a water potential near zero and grows progressively more slowly at lower potentials down to about −80 bar.[11,14] In non-sterile soil hyphal growth continues only if the water potential is less than −8 to −10 bar. At higher potentials chlamydospores germinate, but the germ tubes lyse or form further dormant chlamydospores. This behaviour in non-sterile soil is attributed to the fact that bacterial activity becomes less as the soil dries and nearly ceases at −10 to −15 bar.[12]

Extreme xerophytism is shown by the fungus *Xeromyces bisporus,* which can grow at −693 bar but is completely inhibited at higher water potentials.[42,48] Its intolerance of high water potentials may explain why *Xeromyces* has never been isolated from soil. However, as pointed out by Griffin,[20] some isolates of the *Aspergillus glaucus* group are also inhibited by high water potentials, but do occur in soil. The ability of most soil fungi to grow in soil at much lower water potentials than will support bacterial growth must confer great advantages on them in dry soil. Griffin argues that the reason why bacteria generally cease to grow at potentials much lower than −15 bar is that they then lack motility and as a result are deprived of nutrients. Not only are bacteria unable to move to new nutrient sites in drier soils, but also diffusion paths of soluble nutrients from source to bacterial cells become increasingly long and narrow. Fungi, on the other hand, because of their filamentous structure and mode of growth, can continue to explore and reach nutrients. Translocation of nutrients within hyphae will make possible concentration at points of growth or development.

Discussion of the relative advantages possessed by fungi at low water potentials has meaning only if there is evidence of low water potentials actually occurring in soil. Papendick, Cochran and Woody's study[41] of dryland wheat soil in north-western USA showed that during June the upper 30 cm may become drier than −100 bar owing to the intense evapotranspiration.

It is to be expected that soil water should have a controlling influence on the movement of fungal zoospores. Stolzy *et al.*[52] suggested that *Phytophthora* zoospores require water-filled pores at least 40–60 μm in diameter before they can move through soil. Even when such pores are present, movement through greater distances than a few millimetres probably depends on mass flow of water during and immediately after heavy rainfall. Allen and Newhook[1] have recently used glass capillaries simulating soil pores in the rhizosphere to study the chemotaxis of *Phytophthora* zoospores. They comment that the presence of an ethanol gradient in a capillary tube may allow of normal movement along it that would be impossible in the absence of a gradient, because of the frequency of collision with the walls.

It is easy to understand why sheathing mycorrhizas and their fungi have been studied apart from the ecology of other soil organisms. Discussion of how sheathing mycorrhizal fungi may have evolved their mutualistic symbiotic mode of life will help to relate them to free-living soil fungi. The view most favoured is that mycorrhizal fungi have evolved from aggressive plant parasites.[23, 34] Such evolution must have involved progressive loss, or delicate control, of cytolytic enzymes and a gradual prolongation of a biotrophic rather than a necrotrophic phase. This hypothesis, although attractive, is rather difficult to accept for most sheathing mycorrhizal fungi. None of the predominant genera forming sheathing mycorrhizal associations contains parasitic species or is closely related to other genera containing parasites. On the contrary, most genera containing mycorrhizal species include some more or less saprophytic species or are taxonomically close to saprophytic genera. Furthermore, mycorrhizal species show a range in their degree of dependence on their autotrophic partner. Thus, most members of genera such as *Lactarius* and *Russula* appear to be ecologically obligate symbiotic biotrophs[33] and it is certain that no mycorrhizal species of *Russula* forms sporophores away from a mycorrhizal partner. They are probably unable to grow significantly in natural conditions unless in close contact with a suitable tree root. Singer's observation[49] that primitive members of the genera *Lactarius* and *Russula* are non-mycorrhizal, while higher forms are mycorrhizal and even specialised, suggests evolution from saprotrophy to biotrophy within these genera.

The contention that many sheathing mycorrhizal fungi are ecologically obligate biotrophs is difficult to prove. It is surmised from the difficulty of culturing some species and total failure to culture others. As noted earlier, it has been estimated that as many as 40% of ectomycorrhizal fungi cannot be cultured on known laboratory media. However, species such as *Boletus bovinus, Boletus badius, Boletus subtomentosus, Laccaria laccata* and *Laccaria amethystina* are less wholly dependent on mycorrhizal associations.[23, 34, 36]

Paxillus involutus and possibly some species of *Hebeloma*[22] are examples of fungi that may be classed as ecologically facultative symbionts and whose nutrition is biotrophic in the symbiotic mode, but otherwise saprophytic.[33] *Paxillus involutus* is a widespread woodland species that grows and fruits in ectotrophic and anectotrophic woodlands,[50] and may even form sporophores on rotting logs on the woodland floor. It is probably a consequence of its saprophytic tendencies that this species is very easily isolated from mycorrhizal roots. It is also relatively unspecific in forming mycorrhizal associations. Mycorrhizal Gasteromycetes, such as *Scleroderma aurantium,* and the Ascomycete *Cenococcum graniforme*[56] are also sheathing mycorrhizal fungi that can grow independently of symbiosis.

The following sequence of events may have led to the evolution of obligately symbiotic biotrophs from obligately free-living saprophytic Hymenomycetes.

(1) Adoption of the rhizoplane as the habitat of part of the thallus.
(2) Loss of cellulolytic and lignolytic enzymes, if previously possessed, or development of a delicate mechanism to control their action (some sheathing mycorrhizal fungi are capable of, usually limited, cellulose utilisation).
(3) Penetration between root cortical cells and accompanying increased development on the root surface (loss of host root hairs may have occurred during this phase of evolution).

(4) Loss of ability to synthesise certain growth-regulatory compounds and concomitant partial loss of competitive saprophytic ability.

Loss of the ability to reproduce sexually, i.e. to produce sporophores, in the saprophytic mode may have occurred some time before total loss of ability to grow saprophytically. It is possible that some sheathing mycorrhizal fungi may produce sporophores rarely or not at all.[31] Those basidiomycetes that can no longer form sporophores are likely to be among the most highly evolved mycorrhizal symbionts.

A major feature that distinguishes mycorrhizal fungi from most other root-infecting biotrophs and from necrotrophs is that part of the thallus remains active in the soil following infection. The thallus is in fact differentiated into two parts having nutritionally very different functions. Extending outwards through the soil from the mycorrhizal surface is a system of individual and aggregated hyphae. Mycelial 'strands' usually predominate. These are simple branching aggregations of hyphae quite distinct from highly organised rhizomorphs;[6] they may extend centimetres from the root and themselves bear individual hyphae that grow into the soil. This system of 'strands' and hyphae effectively explores a much larger volume of soil than non-mycorrhizal roots. Its role in the woodland ecosystem may be compared in some respects to that of plant roots. It absorbs water and mineral nutrients in competition with other soil organisms, but does not usually compete for organic nutrients. It differs from plant roots in its greater exploratory powers, its possible lack of exudates usable by saprophytes and its frequent production of antibiotics.

A unique property of the soil mycelium of at least some sheathing mycorrhizal fungi is the ability to transfer soluble carbon compounds from one plant to another of the same or different species. Thus, Björkman[3] demonstrated the movement of ^{14}C from spruce trees to plants of *Monotropa* through the hyphae of a shared mycorrhizal fungus. In addition to its absorptive and translocatory functions, the soil mycelium initiates new root infections and produces sporophores.

As already pointed out, it is doubtful whether the more specialised sheathing mycorrhizal fungi can grow in the absence of their tree partner. Survival without growth may be possible and could be important—for example, in forest tree nurseries. Potential survival propagules are sclerotia, chlamydospores, oidia and basidiospores.

175

Lamb and Richards[32] studied the survival potential of some of these propagules. Chlamydospores of three unidentified mycorrhizal fungi from pine roots were found to lose viability quickly, even at optimal humidity, and also to have poor heat tolerance.

The second part of the sheathing mycorrhizal fungus thallus is closely associated with the host root. It comprises a more or less well-developed sheath, ranging in structure from simple prosenchyma to compacted pseudoparenchyma or synechyma,[63] and hyphae penetrating inter- and intracellularly from the sheath into the cortex. The sheath protects the root and serves as a temporary store for nutrients such as phosphates, while the intercellular Hartig net and the intracellular haustoria provide sites for nutrient exchange between fungus and root.

Interactions between the soil mycelium of sheathing mycorrhizal fungi and soil organisms have been little studied. There has, however, been interest in the rhizosphere of sheathing mycorrhizas, work on which was recently reviewed by Rambelli.[44] In most cases where critical comparison has been made, a greater number and density of micro-organisms have been found in rhizospheres of mycorrhizal than of non-mycorrhizal roots.

Our knowledge of vesicular–arbuscular (VA) mycorrhizal fungi as soil organisms is even more scanty than that of sheathing mycorrhizal fungi. Species of *Endogone,* the major VA fungi, neither form conspicuous fruiting bodies nor can be isolated and cultured from soil, as they appear to be more or less ecologically obligate symbionts. Fortunately, they have features that have made it possible to learn something of their distribution in soil. These are characteristic mycelium and equally characteristic large spores. Not only can *Endogone* be recognised from these structures, but species have been based mainly or entirely on spore morphology. Spores of *Endogone* can be extracted from soil quantitatively by wet sieving, to study numbers and distribution of spore types in different soils and the effects of different crops and treatments. Numbers of spores extractable from a soil do not necessarily give a good guide to the amount of infection of roots in that soil; instances have been recorded when, despite heavy root infection, no spores could be recovered. Work on these aspects of *Endogone* ecology has been reviewed by Mosse.[39]

Probably because of their wide host range, most 'species' (or 'spore types') of *Endogone* appear to have a very wide distribution. An example, cited by Mosse, is the distribution of the type with easily identifiable honey-coloured spores. This has been found in Scotland,

England, Australia, New Zealand, Pakistan, South Africa, the USA and Brazil. As mycorrhizal *Endogone* spp. do not produce spores that become airborne, distribution of species may have depended on the transport of plants with infected roots or soil. It would be interesting to study the *Endogone* types present on islands to which plants have never been imported by man. In such isolated situations unique species may have evolved. Reports of VA fungi in the roots of fossil plants such as *Rhynia* and *Asteroxylon*[29] suggest a very ancient origin for these fungi. This may explain their wide host range and geographical distribution. Our knowledge of the relationship of VA fungi to free-living species is too meagre for speculation on the phylogeny of these fungi to be profitable.

While there is evidence, as for sheathing mycorrhizal fungi, that movement of carbon compounds from the host plant into the external mycelium of VA fungi occurs, little is known of the ability of the soil mycelium to take up organic compounds. Experiments with two-membered cultures, i.e. host plant plus *Endogone,* have shown a fungal response to glycerol or inositol added to the medium. It remains uncertain, however, whether such compounds are taken up directly by the mycelium or by the roots with subsequent alteration in the plant's metabolism with maybe an indirect effect on the fungus. Equally, whether uptake of organic substances by VA fungi occurs only through the intracellular arbuscles or whether the external mycelium of *Endogone* competes with soil heterotrophs for organic compounds has not been established. On the other hand, competition for phosphates or other mineral nutrients is probably active and ecologically significant. Some interactions between *Endogone* and other root-infecting organisms have been observed, but these have probably occurred either within the root or at the root surface.

It has been possible in this chapter to discuss only certain aspects of the ecology of soil fungi. Some of the problems of assessing fungal biomass and activity have been described. Not the least of these is due to our relative ignorance of the behaviour in soil of mycorrhizal fungi, particularly those that are obligate or partly obligate symbionts. Intensive study of these fungi in soil should be rewarding and fill many of the gaps in our knowledge of fungal ecology.

REFERENCES

1. ALLEN, R. N. and NEWHOOK, F. J. 'Chemotaxis of zoospores of *Phytophthora cinnamomi* to ethanol in capillaries of spore dimensions, *Transactions of the British Mycological Society,* **61**, 287–302 (1973)

Soil Fungi

2. AUSMUS, B. S. The use of the ATP assay in terrestrial decomposition studies, in *Modern Methods in the Study of Microbial Ecology* (edited by T. Rosswall): *Bulletins from the Ecological Research Committee* (Stockholm), **17**, 223–234 (1973)
3. BJÖRKMAN, E. *Monotropa hypopitys* L. an epiparasite on tree roots, *Physiologia Plantarum,* **13**, 308–329 (1960)
4. BLACKWELL, E. M. The life history of *Phytophthora cactorum, Transactions of the British Mycological Society,* **26**, 71–89 (1943)
5. BOHLOOL, B. B. and SCHMIDT, E. L. A fluorescent antibody technique for determination of growth rates of bacteria in soil, in *Modern Methods in the Study of Microbial Ecology* (edited by T. Rosswall): *Bulletins from the Ecological Research Committee* (Stockholm), **17**, 336–338 (1973)
6. BOWEN, G. D. mineral nutrition of ectomycorrhizae, in *Ectomycorrhizae—their Ecology and Physiology* (edited by G. C. Marks and T. T. Kozlowski). Academic Press, New York (1973)
7. BROCK, T. D. Bacterial growth rate in the sea: direct analysis by thymidine autoradiography, *Science,* **155**, 81–83 (1967)
8. BROCK, T. D. and BROCK, M. L. Autoradiography as a tool in microbial ecology, *Nature,* **209**, 734–736 (1966)
9. BROCK, T. D. and BROCK, M. L. The application of micro-autoradiographic techniques to ecological studies, *Mitteilungen Internationaler Vereinigung für Theoretische und Angewandte Limnologie,* **15**, 1–29 (1968)
10. CHEN, A. W. and GRIFFIN, D. M. Soil physical factors and the ecology of fungi. V. Further studies in relatively dry soils, *Transactions of the British Mycological Society,* **49**, 419–426 (1966)
11. COOK, R. J. and PAPENDICK, R. I. Soil water potential as a factor in the ecology of *Fusarium roseum* f. sp. *cerealis* 'Culmorum', *Plant and Soil,* **32**, 131–145 (1970)
12. COOK, R. J. and PAPENDICK, R. I. Effect of soil water on microbial growth, antagonism and nutrient availability in relation to soil-borne fungal diseases of plants, in *Root Diseases and Soil-Borne Pathogens* (edited by T. A. Tousson, R. V. Bega and P. E. Nelson). University of California Press, Berkeley (1971)
13. COOK, R. J. and PAPENDICK, R. I. Influence of water potential of soils and plants on root disease, *Annual Review of Phytopathology,* **10**, 349–374 (1972)
14. COOK, R. J., PAPENDICK, R. I. and GRIFFIN, D. M. Growth of two root rot fungi as affected by osmotic and matrix water potentials, *Soil Science Society of America Proceedings,* **36**, 78–82 (1972)
15. DANIELSON, R. M. and DAVEY, C. B. The abundance of *Trichoderma* propagules and the distribution of species in forest soils, *Soil Biology and Biochemistry,* **5**, 485–494 (1973)
16. DOBBS, C. G. and HINSON, W. H. A widespread fungistasis in soils, *Nature,* **207**, 1354–1356 (1953)
17. GRAY, T. R. G. and WILLIAMS, S. T. *Soil Microorganisms.* Oliver and Boyd, Edinburgh (1971)
18. GRIFFIN, D. M. Soil moisture and the ecology of soil fungi, *Biological Reviews,* **38**, 141–166 (1963)
19. GRIFFIN, D. M. Soil water in the ecology of fungi, *Annual Review of Phytopathology,* **7**, 289–310 (1969)
20. GRIFFIN, D. M. *Ecology of Soil Fungi.* Chapman and Hall, London (1972)
21. HACSKAYLO, E. Carbohydrate physiology of ectomycorrhizae, in *Ectomycorrhizae: their Ecology and Physiology* (edited by G. C. Marks and T. T. Kozlowski). Academic Press, New York (1973)
22. HACSKAYLO, E. and BRUCHET, G. *Hebelomas* as mycorrhizal fungi, *Bulletin of the Torrey Botanical Club,* **99**, 17–20 (1972)

178

23. HARLEY, J. L. *The Biology of Mycorrhiza,* 2nd edition. Leonard Hill, London (1969)
24. HOLM-HANSEN, O. The use of ATP determinations in ecological studies, in *Modern Methods in the Study of Microbial Ecology* (edited by T. Rosswall): *Bulletins from the Ecological Research Committee* (Stockholm), **17**, 215–222 (1973)
25. HORA, T. S. and BAKER, R. Extraction of a volatile factor from soil inducing fungistasis, *Phytopathology,* **62**, 1475–1476 (1972)
26. HORA, T. S. and BAKER, R. Soil fungistasis: microflora producing a volatile inhibitor, *Transactions of the British Mycological Society,* **59**, 491–500 (1972)
27. JACKSON, R. M. Studies of fungi in pasture soils. II. Fungi associated with plant debris and fungal hyphae in soil, *New Zealand Journal of Agricultural Research,* **8**, 865–877 (1965)
28. JONES, P. C. T. and MOLLISON, J. E. A technique for the quantitative estimation of activity of fungi in soil, *Journal of General Microbiology,* **2**, 54–69 (1948)
29. KIDSTON, R. and LANG, W. H. On Old Red Sandstone plants showing structure from the Rhyni Chert bed, Aberdeenshire, Part V, *Transactions of the Royal Society of Edinburgh,* **52**, 855–902 (1921)
30. KNIGHT, R. A. Investigations into the germination of conidia of *Cochliobolus sativus* with particular reference to soil fungistasis, PhD thesis, University of Surrey (1970)
31. LAMB, R. J. and RICHARDS, B. N. Some mycorrhizal fungi of *Pinus radiata* and *P. elliottii* var. *elliottii* in Australia, *Transactions of the British Mycological Society,* **54**, 371–378 (1970)
32. LAMB, R. J. and RICHARDS, B. N. Survival potential of sexual and asexual spores of ectomycorrhizal fungi, *Transactions of the British Mycological Society,* **62**, 181–191 (1974)
33. LEWIS, D. H. Concepts in fungal nutrition and the origin of biotrophy, *Biological Reviews,* **48**, 261–278 (1973)
34. LEWIS, D. H. Microorganisms and plants: the evolution of parasitism and mutualism, in *Evolution in the Microbial World* (edited by M. J. Carlile and J. J. Skehel) Symposium 24 of The Society for General Microbiology. Cambridge University Press (1974)
35. LOCKWOOD, J. L. Soil fungistasis, *Annual Review of Phytopathology,* **2**, 341–362 (1964)
36. MEYER, F. H. Distribution of ectomycorrhizae in native and man-made forests, in *Ectomycorrhizae: their Ecology and Physiology* (edited by G. C. Marks and T. T. Kozlowski). Academic Press, New York (1973)
37. MILLAR, W. N. and CASIDA, L. E., Jr. Evidence for muramic acid in soil, *Canadian Journal of Microbiology,* **16**, 299–304 (1970)
38. MIRCETICH, S. M. and ZENTMYER, G. A. Production of oospores and chlamydospores of *Phytophthora cinnamomi* in roots and soil, *Phytopathology,* **56**, 1076–1078 (1966)
39. MOSSE, B. Advances in the study of vesicular–arbuscular mycorrhiza, *Annual Review of Phytopathology,* **11**, 171–196 (1973)
40. NAGEL-DE BOOIS, H. M. Preliminary estimation of production of fungal mycelium in forest soil layers, *Proceedings of the 4th Colloquium of the Soil.* 2nd Communication of the International Soil Science Society (1972)
41. PAPENDICK, R. I., COCHRAN, V. L. and WOODY, W. M. Soil water potential and water content profiles with wheat under low spring and summer rainfall, *Agronomy Journal,* **63**, 731–734 (1971)
42. PITT, J. I. and CHRISTIAN, J. H. B. Water relations of xerophilic fungi isolated from prunes, *Applied Microbiology,* **16**, 1853–1858 (1968)
43. POSTGATE, J. R. The viability of very slow-growing populations: a model for the natural ecosystem, in *Modern Methods in the Study of Microbial Ecology* (edited by T. Russwall): *Bulletins from the Ecological Research Committee* (Stockholm), **17**, 287–292 (1973)

179

44. RAMBELLI, A. The rhizosphere of mycorrhizae, in *Ectomycorrhizae their Ecology and Physiology* (edited by G. C. Marks and T. T. Kozlowski). Academic Press, New York (1973)

45. REEVES, R. J. A study of the biology and ecology of *Phytophthora cinnamomi* Rands, PhD thesis, University of Surrey (1972)

46. REEVES, R. J. Behaviour of *Phytophthora cinnamomi* Rands in different soils and water regimes, *Soil Biology and Biochemistry,* 7, 19–24 (1975)

47. REEVES, R. J. and JACKSON, R. M. Induction of *Phytophthora cinnamomi* oospores in soil by *Trichoderma viride, Transactions of the British Mycological Society,* 59, 156–159 (1972)

48. SCOTT, W. J. Water relations of food spoilage microorganisms, *Advances in Food Research,* 7, 83–127 (1957)

49. SINGER, J. *The Agaricales in Modern Taxonomy.* Cramer, Weinheim (1962)

50. SINGER, R. Forest mycology and forest communities in South America. II. Mycorrhiza sociology and fungus succession in the *Nothofagus dombeyi–Austrocedrus chilensis* woods of Patagonia, in *Mycorrhizae* (edited by E. Hacskaylo). United States Department of Agriculture Washington, DC (1970)

51. SMITH, A. M. Ethylene as a cause of fungistasis, *Nature,* 246, 311–313 (1973)

52. STOLZY, L. H., LETEY, J., KLOTZ, L. J. and LABANAUSKAS, C. K. Water and areation as factors in root decay of *Citrus sinensis, Phytopathology,* 55, 270–275 (1965)

53. SUSSMAN, A. S. Dormancy and spore germination, in *The Fungi,* Volume 2 (edited by G. C. Ainsworth and A. S. Sussman). Academic Press, New York (1966)

54. SWIFT, M. J. The estimation of mycelial biomass by determination of the hexosamine content of wood tissue decayed by fungi, *Soil Biology and Biochemistry,* 5, 321–332 (1973)

55. THOMAS, A., NICHOLAS, D. P. and PARKINSON, D. Modifications of the agar film technique for assaying lengths of mycelium in soil, *Nature,* 205, 105 (1965)

56. TRAPPE, J. M. Mycorrhizae-forming Ascomycetes, in *Mycorrhizae* (edited by E. Hacskaylo). United States Department of Agriculture, Washington DC (1971)

57. WAID, J. S., PRESTON, K. J. and HARRIS, P. J. A method to detect metabolically active microorganisms in leaf litter habitats, *Soil Biology and Biochemistry,* 3, 235–241 (1971)

58. WAID, J. S. and WOODMAN, M. J. A method of estimating hyphal activity in soil, *Pedologie* (Ghent), 7, Numéro Spécial (Symposium des Méthodes et Microbiologie du Sol), 155–158 (1957)

59. WAKSMAN, S. A. Is there any fungus flora of the soil?, *Soil Science,* 3, 567–589 (1917)

60. WARCUP, J. H. On the origin of colonies of fungi developing on soil dilution plates, *Transactions of the British Mycological Society,* 38, 298–301 (1955)

61. WARCUP, J. H. Studies on the occurrence and activity of fungi in a wheatfield soil, *Transactions of the British Mycological Society,* 40, 237–262 (1957)

62. WARCUP, J. H. Methods for isolation and estimation of activity of fungi in soil, in *The Ecology of Soil Fungi* (edited by D. Parkinson and J. S. Waid). Liverpool University Press (1960)

63. ZAK, B. Classification of ectomycorrhizae, in *Ectomycorrhizae: their Ecology and Physiology* (edited by G. C. Marks and T. T. Kozlowski). Academic Press, New York (1973)

64. ZAK, B. and MARX, D. H. Isolation of mycorrhizal fungi from roots of individual slash pines, *Forest Science,* 10, 214–222 (1964)

10 Microbial Degradation of Plant Protection Chemicals

N. WALKER

10.1 Introduction

The widespread use of synthetic organic substances for combating insect pests, weeds, or plant-pathogenic fungi or nematodes and so improving crop yields may be regarded essentially as a post-war phenomen. It is a development that has proceeded at an astonishing rate during the last 25 years and now the range of such plant-protection chemicals is very extensive. This development and the progressively increasing use of chemical fertilisers are by far the biggest factors determining the efficiency and yields of modern agriculture. The fate of these substances in the environment has now become a matter of much public concern.

The discovery of the insecticidal properties of dichloro-diphenyl-trichloroethane (DDT) in 1939 by the Geigy Chemical Company together with the recognition of the value of 2,4-di-chlorophenoxyacetic acid (2,4-D) as a selective herbicide against dicotyledonous weeds during the course of World War II by, among others, Nutman, Thornton and Quastel[24] working at Rothamsted marked the beginning of this modern era of plant protection. Incidentally, these authors also observed the limited persistence of 2,4-D in unsterilised soil and so initiated what has now become an extensive range of studies on the microbial degradation of herbicides and pesticides.

Moreover, the above discoveries serve to typify two diverging aspects of these investigations. Firstly, DDT, which has proved to be an extremely stable, effective and persistent insecticide, has given rise to much public concern over the pollution of the environment due to its widespread use and consequent dangers to wild life. This concern has been stimulated by publications such as Rachel Carson's *Silent Spring*[9] and, in consequence, researches into pesticide residues in soils plants and animals have greatly multiplied. On the other hand, 2,4-D is an example of an effective herbicide also used on a massive scale, but of only limited persistence in soil because it is readily degraded by many soil micro-organisms.[25] Thus, the metabolism of 2,4-D and related compounds by many microbial species, the metabolic pathways involved and the relevant enzymic reactions have been investigated in much detail.[12, 14] These investigations have, at the same time, shed light on the very varied catabolic potentialities of the microbial world.[25] Scientific work on problems of this kind extending into different aspects of environmental pollution has become so extensive as to justify publication of several specialist journals expressly devoted to such subjects; in addition, the numerous books, articles and reviews now extant are ample evidence of the vast amount of work accomplished. It is clear, therefore, that in this short essay it is impossible to do more than outline a small selection of such studies. The list of publications and reviews on p. 192 will provide a guide to the extensive recent work. Here an attempt will be made to discuss some general principles underlying investigations of the role of soil micro-organisms in degrading plant-protection and related compounds and a few examples of investigations, to some of which the author has contributed, will be given.

10.2 Stability and persistence of synthetic compounds in soil

Soils are complex mixtures of weathered minerals and plant and animal organic debris in different stages of decay, and they contain a wide variety of plant, animal and microscopic life. Soils are exposed to many physical, chemical or climatic influences, including varying temperatures, wetting and drying, radiations, wind, erosion, and so on. Consequently, chemical substances added to soil are subjected similarly to a variety of chemical, physical and biological forces, and they may react or be degraded in many ways. Here only microbial effects will be considered, but one should be constantly aware that substances reaching soil may be dispersed or decomposed by mechanisms other than biological ones, such as leaching, volatilisation, chemical oxidations and reactions and photochemical reactions, as well as by plant and animal processes.

Only very stable, insoluble compounds can be expected to remain in soil for long periods unaffected by such environmental influences and, of course, there are several substances in this category which are or have been used for agricultural purposes. Many of these are chlorinated hydrocarbons, e.g. DDT (see reference 26), aldrin, dieldrin, heptachlor, etc.,[22] but other substances—for example, some triazine herbicides (simazine)—are equally stable and so persistent that any slow microbial attack on them can scarcely be regarded as normal metabolism. Molecules of this kind have been termed 'recalcitrant' by Alexander[1] and, in general, they are quite resistant to microbiological decomposition. So, in spite of the versatile biochemical activities of the microbial world, many organic molecules are completely resistant to their attack. The persistence in soil of substances of this kind has sometimes been assessed by estimations of their 'half-life'—that is, the time needed for the initial concentration of the substance to decline to half; some of the stable chlorinated hydrocarbon insecticides have half-lives of several years (see *Environmental Pollution by Pesticides,* listed on p. 192).

10.3 Utilisation of aromatic compounds by micro-organisms

Many plant-protection chemicals are aromatic in nature—that is, are derived from benzene or other similar carbocyclic hydrocarbons.

The stability of carbocyclic ring compounds has been recognised by chemists ever since the development of the coal tar dyestuff and chemical industry in the second half of the nineteenth century. Nevertheless, although the degradation of aromatic compounds such as phenol or toluene by bacteria has been known since the beginning of the present century, the ability of micro-organisms to metabolise the benzene ring with such ease in comparison with the drastic reagents needed by the chemist for cleavage of the aromatic nucleus still remains impressive. It is instructive to note, however, that concentration is a factor in determining the growth of bacteria on, say, phenols as their sole carbon source. Few bacteria can assimilate phenol if the concentration exceeds 0.1% and most require the concentration to be less than 0.05%. On the other hand, phenol at concentrations greater than 0.1% is bactericidal, and its effectiveness as an antiseptic is increased by chlorine substitution, as evidenced by an increased Rideal–Walker coefficient.

Correspondingly, concentrations of mono- and di-chlorophenols in aqueous solution, at which they may become susceptible to bacterial degradation, are very much smaller than for phenol itself. Indeed, 2,4-dichlorophenol is only metabolised by bacteria at concentrations less than 0.005%.

Perhaps the first detailed biochemical study of the bacterial metabolism of phenol and benzoic acid was the publication of Evans[11] in 1947, which emerged from earlier work by Happold on phenol and thiocyanate decompositions in coke oven effluents in the 1930s. Evans[11] established that *Vibrio* 01 dissimilated both phenol and benzoic acid via catechol, although the hypothesis was entertained for a time that benzoic acid might have been metabolised through a hydroxybenzoate. After ring fission, an unidentified keto acid giving a positive Rothera reaction was detected. Later work by various other investigators eventually revealed two metabolic pathways. One is through *cis,cis*-muconic acid, a muconolactone, β-keto-adipic acid, succinic, acetic and formic acids, finally to CO_2. Another pathway proceeds via hydroxy-muconic acid semi-aldehyde. These two fission mechanisms are due to two different catechol oxidases, one causing fission between carbon atoms 1 and 2 ('ortho' fission) and the other fission between carbon atoms 2 and 3 ('meta' fission). These mechanisms have been discussed by, among other authors, Stanier, Palleroni and Doudoroff,[31] and much work has been done on the genetic control and induction of these catechol fission mechanisms in various bacterial species. More recently evidence for the anaerobic dissimilation of benzoate by *Rhodopseudomonas Palustris* has been obtained,[10, 16] although the anaerobic utilisation of benzoate by methanogenic microbes has been known for some time.[32] However, decomposition of hydrocarbons and aromatic substances by micro-organisms is usually aerobic, and the most general mode of attack involves a hydroxylation, with *o*-dihydroxy compounds as the key intermediates prior to ring fission.

Other carbocyclic compounds—for example, naphthalene, phenanthrene, diphenyl derivatives—are also susceptible to microbial oxidation, and the possible degradation of such compounds and their derivatives has been much studied. For example, Alexander and Lustigmann[3] used a rapid screening method, based on the loss of characteristic UV absorption when the benzene ring is disrupted, to detect the biological decomposition of such compounds in soil. Aromatic compounds possessing electron-donating substituents, such as OH, alkyl or NH_2 groups, are

184

generally more prone to oxidative metabolism than those with electron-attracting substituents, such as NO_2, halogen or SO_3H.[2] Analogously, in the reaction of peroxidase with monosubstituted anilines increased reactivity is shown by substituents with increasing electron density.[6] Knowledge of the susceptibility to microbial attack on the part of parent aromatic compounds provides useful basic information for investigation of more complex pesticidal substances.

10.4 Microbiological degradation of selected herbicides and pesticides

The earliest studies on microbial herbicide decomposition were of 2,4-dichlorophenoxyacetic acid (2,4-D), 4-chloro-2-methyl-phenoxyacetic acid (MCPA) and related compounds,[4] and this work is of particular relevance because it illustrates certain general techniques and methods of investigation. Audus[4] studied the dynamics of 2,4-D and MCPA disappearance from soil and demonstrated the biological nature of the processes. When a dilute solution of 2,4-D was continuously percolated through a column of soil crumbs by recycling, the time course of the decline in concentration showed three phases. Initially, a slight fall in 2,4-D concentration was explained by adsorption of a small amount on to soil. Then followed a fairly long period with scarcely any change in concentration, and this indicated a lag period during which, either as a result of mutations or by adaptation, a population of micro-organisms able to metabolise the 2,4-D began to develop. The third phase was a rapid logarithmic breakdown of 2,4-D, coincident with the growth of the 2,4-D-adapted microbial population, and this ceased only when the 2,4-D supply was exhausted. A second dose of 2,4-D was immediately decomposed at a similar rate in the course of two or three days, no adsorption or lag phase being then noted. Utilisation of 2,4-D by the adapted micro-organisms was influenced only by such factors as affect aerobic bacterial growth generally, such as temperature, pH, moisture conditions, aeration and substrate or other nutrient supplies. Whether microbes acquire the capacity to utilise some foreign exotic substance through mutations or due to induced enzymes is still an unresolved problem. Evidence that herbicide or pesticide degradation in soil is microbial is also afforded by a study of effects of inhibitors (sodium azide, cyanide, chloroform), pasteurisation or sterilisation on the disappearance of the substance.

185

Approaches of this kind have been frequently adopted in later work on the fate of pesticides in soils. Final proof that decomposition of a particular plant-protection chemical is microbiological is usually obtained by isolating the effective organism and demonstrating that this can degrade the chemical in pure cultures or enzymically.[12, 14] When the chemical can be utilised by the organism as a nutrient carbon source, then decomposition generally goes to completion. If addition of a pesticide to soil results in its eventual microbiological degradation and, in consequence, growth of the adapted microbial population, then subsequent additions of the pesticide are degraded more rapidly than the first and so remain in soil for a shorter time. Some examples of microbial degradations of various pesticides will now be cited. Parathion, the diethyl ester of *p*-nitro-phenylphosphorothioate and a typical organophosphorus insecticide, is decomposed by various soil micro-organisms. Its insecticidal activity depends on its enzymic conversion to the oxygen analogue, Paroxon. Under anaerobic conditions,[28, 29] as in submerged rice soils, it may be hydrolysed by microbes to *p*-nitrophenol and inorganic phosphate or, in presence of carbohydrates, reduced to *p*-aminoparathion. *p*-Nitrophenol is metabolised by some pseudomonads[27] which cleave the nitro group, giving free nitrous acid and quinol. The latter may be further metabolised by other pseudomonad species. Up to now there is no evidence that the nitro group of Parathion is similarly hydrolysed. The degradation of Parathion, therefore, is rather complicated, and the particular decomposition depends on the organisms and on soil conditions.

Other organophosphorus insecticides are subject to microbial decomposition: for example, Malathion[23] may be degraded to various intermediates by *Rhizobium* spp. or by *Trichoderme viride*. These insecticides, in general, are much less persistent than the earlier, very stable organochlorine compounds.[22]

Phenylcarbamate, phenylurea and acylanilide compounds are now used for plant-protection purposes, either as fungicides, herbicides or insecticides, and nearly all of them are derived from different substituted anilines. Most are known to be degradable in soil by micro-organisms and usually the parent anilines are liberated even though the degradative pathway may vary.[20, 35] The microbial degradation of aniline compounds is an important aspect of the fate of such herbicides in soil. Hill *et al.*[17] found soil organisms able to utilise Monuron (3-(*p*-chlorophenyl)-1,1-dimethylurea) as a carbon source, and, later, other organisms including fungi were also shown

186

to degrade Monuron, Linuron, a chloro derivative of Monuron, was also shown to be degraded with liberation of 3,4-dichloroaniline. Geissbühler *et al.*[15] studied the more complex substance Chloroxuron, and found that it was degraded in a step-wise manner to yield 4-chlorophenoxy-aniline. It is known that aniline and some of its chloro derivatives are metabolised by certain bacteria, but it has also been observed that mono- or di-chloroanilines may be converted in soil non-biologically to chloro-substituted azobenzenes. Such by-products, however, are not normally detectable in soil receiving the corresponding phenylurea or phenylcarbamate herbicides at normal rates of application. Di-, tri-, or tetra-chloroazobenzenes have been detected in soil following excessive dose rates of the herbicides in question. Kaufman and Blake[20] observed that Propham (isopropylcarbanilate), Propanil (dichloropropionanilide), Solan (3'-chloro-2-methyl-*p*-valero-toluidide) and Swep (methyl 3,4-dichlorocarbanilate) were degraded in various soil enrichments cultures as well as by pure cultures of micro-organisms isolated therefrom. In each case the herbicide disappeared and the parent aniline and ionic chlorine were liberated. The responsible organisms included pseudomonads, *Achromobacter* spp. and several fungi (*Aspergillus, Penicillium* and *Fusarium* spp.). Kaufman and Kearney[21] obtained enzyme preparations from a *Pseudomonas* sp. which hydrolysed the ester linkage of phenylcarbamates, giving the alcohol moiety and the parent aniline. Bachofer, Oltmanns and Lingens[5] described a *Nocardia*-like soil micro-organism which grew on various carboxy-anilide fungicides as well as on phenol, catechol, aniline and protocatechuate. The fungicides *o*-toluanilide and 2,5-dimethyl-furan-3-carboxyanilide and the herbicide isopropyl-*N*-phenyl-carbamate were all metabolised. Wallnöfer and his co-workers[34,35] have studied the decomposition of several phenylurea herbicides by a soil *Bacillus sphaericus* strain. They have shown that *B. sphaericus* hydrolyses the acylamide group, affording the corresponding anilines and acids; 2-methyl- and 2-chloro-benzoic acid anilines and the various N'-methoxyphenylureas, Monolinuron, Linuron, Metabromuron and Maloran are all degraded. The suggested pathway is:

$$R_1 - \langle\!\!\!\bigcirc\!\!\!\rangle - NHCO\ R_3 + H_2O \rightarrow R_1 \langle\!\!\!\bigcirc\!\!\!\rangle NH_2 + R_3COOH$$
$$R_2 \qquad\qquad\qquad R_2$$

R_2 where $R_1, R_2 = Cl$; $R_3 = NH \cdot Alk$, $O \cdot Alk$, alkyl group

The authors also demonstrated the presence of an acylamidase in *B. sphaericus* cell-free extracts.

The microbial degradation of the free radical, contact herbicides, Paraquat and Diquat, is of some interest. These compounds are effective as general herbicides when sprayed on the above-ground parts of plants, but they are rapidly inactivated by adsorption on soil (see reference 8). They act rapidly and can be used as desiccants on potatoes or other crops and also in pasture renovation without the need for ploughing.

A yeast species, *Lipomyces starkeyi*, has been found to degrade Paraquat (1,1'-dimethyl-4,4'-bipyridylium-2A) and to use it as a nitrogen source. Other organisms also degrade it but not so effectively.[8] Burns and Audus[7] in a recent study of the behaviour of Paraquat in soil confirmed this finding with *Lipomyces* and showed that, when adsorbed on soil organic matter, the compound was still degradable by the yeast, but adsorbed on clay minerals the adsorption was irreversible and it was no longer available to microbial attack. Others have found several micro-organisms that can degrade Paraquat; a *Pseudomonas* sp., for example, converted it to 4-carboxyl-1-methylpyridinium chloride. Diquat is reported to be even more easily degraded by soil bacteria. Many other studies of herbicide and pesticide degradation by microbes are described in the literature and the reader is referred to the various monographs listed at the end of this chapter.

10.5 Biochemical considerations; the role of cometabolism in pesticide decomposition

In many instances, whether soil micro-organisms can utilise and so degrade foreign synthetic compounds that happen to reach soil depends on the induction of suitable enzyme systems in the potentially operative micro-organisms in question. When so adapted, such microbes have an ecological advantage over the vast range of other micro-organisms in that they can use the foreign molecule as a carbon and energy source and so outgrow the rest, which exist in soil on marginal supplies of suitable substrates. It follows that an adapted microbial population then proliferates on its new source of carbon and in so doing decomposes the foreign molecule, generally in stages, ultimately to CO_2 and water. Sometimes there are different metabolic pathways by which these substrates are catabolised in different organisms; the number of such pathways, however, is strictly limited. For example, 2,4-D can be degraded by two and possibly three different initial pathways by the dozen or so species of bacteria able to metabolise it. Usually degradations brought about by induced microbial enzyme systems conform to Stanier's simultaneous adaptation hypothesis, and this is often a useful method for identifying intermediates formed in the metabolic pathway.

There is another mechanism, however, by which micro-organisms can cause at least partial degradation of some foreign molecule—namely, by co-oxidation or cometabolism of substances that are analogues of normal substrates. The term was first employed by Jackson Foster, who had observed the limited oxidation of certain aromatic hydrocarbons added to active cultures of *Nocardia* growing on paraffin hydrocarbons. The organism grew well on, say, hexadecane as sole carbon and energy source but could not utilise methylnaphthalene or mesitylene. By adding either of the latter two compounds to hexadecane cultures, oxidation occurred of the two aromatic compounds to carboxylic acids, naphthoic and *p*-isopropylbenzoic acid, respectively. Unpublished work by the author at Rothamsted in 1951 on the possible microbial degradation of chlorobenzoic acids resulted in the isolation of a soil *Corynebacterium* sp., the benzoate-grown cells of which absorbed oxygen in presence of either 3- or 4-chlorobenzoate. At the time the reason for this was not clear but now it can be recognised as an example of cometabolism. Recently there have been many studies of cometabolism of halogenated benzoic acids and phenols, monochlorobenzoates or monochlorophenols being oxidised to a chlorocatechol.[33] Horvath[18, 19] reported the cometabolism of the herbicides 2,4,5-T and 2,3,6-trichlorobenzoic acid by benzoate-adapted bacteria. Focht and Alexander[13] described the cometabolism of some simpler analogues of DDT, and Raymond and Alexander[27] found that a *p*-nitrophenol-grown organism, possibly a *Flavobacterium* sp., co-oxidised the isomeric *m*-nitrophenol to nitro-quinol, although *m*-nitrophenol did not support growth of the organism.

Spokes and Walker[30] found several different genera of soil bacteria which when adapted to growth on phenol or benzoate could then oxidise various chlorophenols or chlorobenzoates to a chloro-catechol or, in one instance, a benzoate-grown *Bacillus* oxidised 3-chlorobenzoate to 3-chloro-2,3-dihydroxybenzoate. But in all these reactions the formation of catechol derivatives means that the product becomes more prone to further oxidation either chemically

189

or biologically. In this way even a partial oxidation is a step in the direction of eventual detoxification of persistent chlorinated aromatic substances.

While the aim of much of the recent work on microbiological pesticide breakdown has been to establish that the pesticide in question is decomposed in soil without any accumulation of possibly toxic intermediates, knowledge of the relative stability of many molecular structures has also been gained. Often the number of substituting chlorine atoms in a compound or the position of the substituents has a decisive influence on the resistance of the compound to microbial attack. Thus, 4-chlorophenoxyacetic acid and 2,4,-dichlorophenoxyacetic acid are readily degraded by bacteria, whereas 3-chloro- or 2,4,5-trichloro-phenoxyacetic acids are very resistant to breakdown. This kind of information may be helpful in assessing the potential stability and, hence, persistence in soil of new compounds. To the soil microbiologist, investigations of the type described also reveal the diverse capacities of micro-organisms for reacting with or metabolising apparently stable and unusual synthetic substances. This field of work is likely to continue to expand as more and more synthetic chemicals find their way into agriculture.

REFERENCES

1. ALEXANDER, M. *Microbial Ecology*, 413–416. Wiley, New York (1971)
2. ALEXANDER, M. and ALEEM, M. I. H. Effect of chemical structure on microbial decomposition of aromatic herbicides, *Journal of Agricultural and Food Chemistry*, **9**, 44–47 (1961)
3. ALEXANDER, M. and LUSTIGMAN, B. K. Effect of chemical structure on microbial degradation of substituted benzenes, *Journal of Agricultural and Food Chemistry*, **14**, 410–413 (1966)
4. AUDUS, L. J. The biological detoxication of hormone herbicides in soil, *Plant and Soil*, **3**, 170–192 (1951)
5. BACHOFER, R., OLTMANNS, O. and LINGENS, F. Isolation and characterisation of a *Nocardia*-like soil bacterium growing on carboxyanilide fungicides, *Archiv für Mikrobiologie*, **90**, 141–149 (1973)
6. BORDELEAU, L. M. and BARTHA, R. Biochemical transformations of herbicide-derived anilines: requirements of molecular configurations, *Canadian Journal of Microbiology*, **18**, 1873–1882 (1972)
7. BURNS, R. G. and AUDUS, L. J. Distribution and breakdown of Paraquat in soil, *Weed Research*, **10**, 49–58 (1970)
8. CALDERBANK, A. The bipyridylium herbicides, *Advances in Pest Control Research*, **8**, 127–235 (1968)
9. CARSON, RACHEL. *Silent Spring*. Hamish Hamilton, London (1963)

10. DUTTON, P. L. and EVANS, W. C. The metabolism of aromatic compounds by *Rhodopseudomonas palustris, Biochemical Journal,* **113,** 525–536 (1969)

11. EVANS, W. C. Oxidation of phenol and benzoic acid·by some soil bacteria, *Biochemical Journal,* **41,** 373–382 (1947)

12. EVANS, W. C., SMITH, B. S. W., FERNLEY, H. N. and DAVIES, J. I. Bacterial metabolism of 2,4-dichlorophenoxyacetate, *Biochemical Journal,* **122,** 543–551 (1971)

13. FOCHT, D. D. and ALEXANDER, M. Aerobic cometabolism of DDT analogues by *Hydrogenomonas* sp., *Journal of Agricultural and Food Chemistry,* **19,** 20–22 (1971)

14. GAUNT, J. K. and EVANS, W. C. Metabolism of 4-chloro-2-methylphenoxyacetate by a soil pseudomonad, *Biochemical Journal,* **122,** 519–526 (1971)

15. GEISSBÜHLER, H., HASELBACH, C., AEBI, H. and EBNER, L. The fate of N'-(4-chlorophenoxy)-phenyl-NN-dimethylurea (C-1983) in soils and plants. III. Breakdown in soils and plants, *Weed Research,* **3,** 277–297 (1963)

16. GUYER, M. and HEGEMAN, G. Evidence for a reductive pathway for the anaerobic metabolism of benzoate, *Journal of Bacteriology,* **99,** 906–907 (1969)

17. HILL, G. D., McGAHEN, J. W., BAKER, H. M., FINNERTY, D. W. and BINGEMAN, C. W. The fate of substituted urea herbicides in agricultural soils, *Agronomy Journal,* **47,** 93–104 (1955)

18. HORVATH, R. S. Microbial cometabolism of 2,4,5-T, *Bulletin of Environmental Contamination and Toxicology,* **5,** 537–541 (1970)

19. HORVATH, R. S. Cometabolism of the herbicide 2,3,6-trichlorobenzoate, *Journal of Agricultural and Food Chemistry,* **19,** 291–293 (1971)

20. KAUFMAN, D. D. and BLAKE, J. Microbial degradation of several acetamide, acylanilide, carbamate, toluidine and urea pesticides, *Soil Biology and Biochemistry,* **5,** 297–308 (1973)

21. KAUFMAN, D. D. and KEARNEY, P. C. Microbial degradation of isopropyl-N-3-chlorophenylcarbamate and 2-chloroethyl-N-3-chlorophenylcarbamate, *Applied Microbiology,* **13,** 443–446 (1965)

22. LICHTENSTEIN, E. P., SCHULZ, K. R., FUHREMANN, T. W. and LIANG, T. T. Degradation of Aldrin and Heptachlor in field soils during a ten-year period, *Journal of Agricultural and Food Chemistry,* **18,** 100–106 (1970)

23. MOSTAFA, I. Y., FAKHR, I. M. I., BAHIG, M. R. E. and EL-ZAWAHRY, Y. A. Metabolism of organophosphorus insecticides. XIII. Degradation of Malathion by *Rhizobium* spp., *Archiv für Mikrobiologie,* **86,** 221–224 (1972)

24. NUTMAN, P. S., THORNTON, H. G. and QUASTEL, J. H. Inhibition of plant growth by 2,4-dichlorophenoxyacetic acid and other plant growth substances, *Nature,* **155,** 498–500 (1945)

25. PAINTER, H. A. Biodegradability, *Proceedings of the Royal Society B,* **185,** 149–158 (1974)

26. PFAENDER, F. K. and ALEXANDER, M. Extensive microbial degradation of DDT *in vitro* and DDT metabolism by natural communities, *Journal of Agricultural and Food Chemistry,* **20,** 842–846 (1972)

27. RAYMOND, D. G. M. and ALEXANDER, M. Microbial metabolism and cometabolism of nitrophenols, *Pesticide Biochemistry and Physiology,* **1,** 123–130 (1971)

28. SETHUNATHAN, N. and YOSHIDA, TOMIO. Parathion degradation in submerged rice soils in the Philippines, *Journal of Agricultural and Food Chemistry,* **21,** 504–506 (1973)

29. SIDDARAMAPPA, R., RAJARAM, K. P. and SETHUNATHAN, N. Degradation of Parathion by bacteria isolated from flooded soil, *Applied Microbiology,* **26,** 846–849 (1973)

30. SPOKES, J. R. and WALKER, N. Chlorophenol and chlorobenzoic acid cometabolism by different genera of soil bacteria, *Archives of Microbiology,* **96,** 125–134 (1974)

31. STANIER, R. Y., PALLERONI, N. J. and DOUDOROFF, M. The aerobic pseudomonads: a taxonomic study, *Journal of General Microbiology,* **43,** 159–271 (1966)

191

32. TARVIN, D. and BUSWELL, A. M. The methane fermentation of organic acids and carbohydrates, *Journal of the American Chemical Society,* **56,** 1751–1755 (1934)
33. WALKER, N. Metabolism of chlorophenols by *Rhodotorula glutinis, Soil Biology and Biochemistry,* **5,** 525–530 (1973)
34. WALLNÖFER, P. The decomposition of urea herbicides by *Bacillus sphaericus* isolated from soil, *Weed Research,* **9,** 333–339 (1969)
35. WALLNÖFER, P. and ENGELHARDT, G. Der Abbau von Phenylamiden durch *Bacillus sphaericus, Archiv für Mikrobiologie,* **80,** 315–323 (1971)

Suggestions for further reading

Degradation of Herbicides (edited by P. C. Kearney and D. D. Kaufman). Marcel Dekker, New York (1969)
Organic Chemicals in the Soil Environment (edited by C. A. I. Goring and J. W. Hamaker). Marcel Dekker, New York (1972)
Environmental Pollution by Pesticides (edited by C. A. Edwards). Plenum Press, London (1973)
HELLING, C. S., KEARNEY, P. C. and ALEXANDER, M. Behaviour of pesticides in soils, *Advances in Agronomy,* **23,** 147–240 (1971)
MENZER, R. E. Biological oxidation and conjugation of pesticidal chemicals, *Residue Reviews,* **48,** 79–116 (1973)
WRIGHT, S. J. L. Degradation of herbicides by soil microorganisms, in *Microbial Aspects of Pollution* (edited by G. Sykes and F. A. Skinner), 233–254. Society for Applied Bacteriology Symposium Series No. 1; Academic Press, London (1971)

11 Effects of Biocidal Treatments on Soil Organisms

D. S. POWLSON

11.1 The sterilisation and partial sterilisation of soil

Most soils contain a large and diverse microbial population. Some of the micro-organisms are inherently resistant to adverse environmental conditions, and soil organic matter can protect organisms from biocidal chemicals by sorbing the biocide, thus decreasing its effective concentration.[49] It is not surprising, therefore, that soil is a difficult material to sterilise. Small quantities of soil can be sterilised by autoclaving or by a sufficient dose of ionising radiation. Steaming, oven-drying, irradiation at lower doses and treatment with certain chemicals often produce near-sterile conditions in soil, and these treatments are referred to as 'partial sterilisation'. This term sometimes refers to less biocidal treatments, such as air-drying, freezing or mechanical disturbance. In this chapter all such treatments are referred to as biocidal treatments but modified as necessary (e.g. strongly biocidal, slightly biocidial). The term 'sterilisation' is used only in its strict sense and the contradictory term 'partial sterilisation' is avoided.

11.2 Control of soil-borne plant pathogens by biocidal treatments

There are some situations in agriculture and horticulture where soil-borne plant pathogens may be controlled by strongly biocidal treatments that affect most sections of the soil population. The oldest of these treatments is heating. The beneficial effects of burning crop remains were known to the Romans[53] and have long been known in India.[102] Heating glasshouse soils with steam or mixtures of steam and air to control plant diseases caused by fungi or nematodes is widely practised.

Volatile chemicals have been used to fumigate soils so as to kill pathogens for over a century. Tietz[119] and Wilhelm[129] describe early work on soil fumigation and the theories developed to explain its favourable effects on plant growth. Carbon disulphide was the first chemical used as a soil fumigant. Those used today include methyl bromide, trichloronitromethane (chloropicrin), dichloropropene (mixed with dichloropropane and C_3 hydrocarbon as 'D-D mixture'), ethylene dibromide, methyl isothiocyanate and formaldehyde. In addition, some compounds are used which decompose in soil to evolve fumigants—e.g. sodium ethylene bisdithiocarbamate (Nabam), which decomposes to carbon disulphide; and sodium N-methyldithiocarbamate (Vapam or metham sodium) and tetrahydro-3,5-dimethyl-2H-1,3,5-thiadiazine-2-thione (Dazomet or Mylone), which both decompose to methyl isothiocyanate. The physical aspects of soil fumigation, including factors affecting the movement of fumigants through soil, have been reviewed by Goring.[49, 50]

Various reviews[39, 49, 68, 85, 87, 129, 132] contain many references to the control of plant diseases by fumigants and the resulting yield increases. Commercially, fumigants are mainly used for high-value crops because of the high cost of treatment, although the cost is less if the fumigant is applied only to the rows and not to the whole field. In some countries (USA, the Netherlands) fumigants are used on a field scale in the production of fruit, vegetables, ornamental flowers and bulbs. They have successfully restored the productivity of soils long under monoculture and considered 'worn out'; examples are citrus replants in California[129] and orchard replants in the Netherlands.[39, 56] In the United Kingdom soil fumigants have been used mainly for glasshouse crops and also in forest nurseries.[4] Recently soil fumigation in the field with D-D, chloropicrin, ethylene dibromide or Telone (mostly 1,3-dichloropropene) has controlled stubby root nematodes (*Trichodorus* spp.) and needle nematodes (*Longidorus* spp.), which cause severe stunting of sugar beet, known as 'docking disorder' (see reference 127). Fumigating the rows where sugar-beet is later sown may control this condition economically.[128]

Experimentally, a biocidal treatment may be used to determine whether a particular abnormality in plant growth is caused by a chemical (e.g. a nutrient deficiency) or a biological mechanism (e.g. a parasitic nematode). When some soil organism is to be studied, it may be necessary to eliminate interference from other organisms by first sterilising the soil.

11.3 Changes in the soil population

The effect of biocidal treatments on soil chemical and biological properties has been reviewed many times. Powlson[91] listed 28 reviews and Goring[50] 26 reviews on soil fumigation alone. Most of the effects were discovered more than 50 years ago and have been often reviewed.[2, 10, 17, 32, 45, 53, 54, 67, 69, 75, 76, 79, 85, 118, 121, 122, 126, 129]

By definition, biocidal treatments kill organisms, so that their numbers in soil are decreased. This section is mainly concerned with the relative killing power of different treatments and the population changes occurring after the initial decrease in numbers.

Fumigants and heat treatments

Strongly biocidal treatments, such as fumigation, steaming or oven-drying followed by re-wetting, all have broadly similar effects on numbers of organisms. The results of Reber[93] (*Figure 11.1*) illustrate these effects in soil fumigated with chloropicrin. *Table 11.1* (after Martin[75]) shows the effects of different fumigants on numbers of bacteria and actinomycetes in soil.

One of the most dramatic effects is on the numbers of bacteria; after an initial decrease, bacterial numbers increase rapidly and greatly exceed those in untreated soil. The time at which the peak in numbers occurs varies with the treatment, the type of soil and the experimental conditions. It usually occurs slightly earlier if the treated soil is inoculated with fresh soil and the numbers attained may then be greater (*Figure 11.1*). The more effective the biocidal treatment, the smaller the population of survivors and the slower the subsequent recolonisation. In reporting the effects of biocidal treatments on numbers of soil organisms, it is important to specify whether the treated soil was kept sterile, inoculated with fresh soil or left open to chance contamination. After first increasing, bacterial numbers decline and gradually approach those in untreated soil but the time required for this varies; e.g. about 14 weeks (*Figure 11.1*) to over a year (*Table 11.1*). There may be several peaks in

195

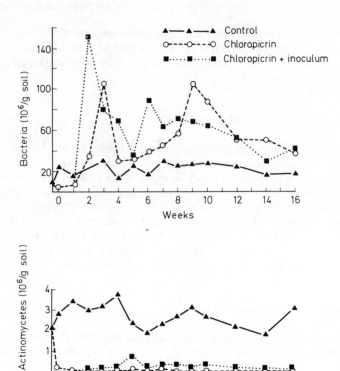

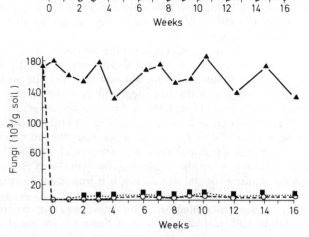

Figure 11.1 Effect of chloropicrin fumigation and inoculation with untreated soil on the numbers of bacteria, actinomycetes and fungi in a soil (from Reber[93])

Table 11.1

EFFECT OF SOIL FUMIGATION ON NUMBERS OF BACTERIA IN YOLO SANDY LOAM
IN 3-GALLON POTS IN THE GREENHOUSE (FROM MARTIN[76])

Treatment*	Numbers† of bacteria plus actinomycetes per g dry soil after incubation, days					
	1	10	20	50	100	250
Check	37	29	23	19	22	9
D-D mixture, 400 lb/acre	22	88	88	48	58	19
D-D mixture, 4000 lb/acre	2	99	95	52	64	61
Chloropicrin, 200 lb/acre	4	96	120	61	60	48
Chloropicrin, 2000 lb/acre	2	100	153	78	70	48
Carbon disulphide, 6000 lb/acre	3	84	91	86	91	72
Propylene oxide, 2000 lb/acre	2	80	154	90	98	60
Ethylene dibromide, 200 lb/acre	36	29	29	21	29	9
Ethylene dibromide, 2000 lb/acre	9	68	60	50	53	23

* Dosages calculated on acre 6-inch basis (2 000 000 lb).
† Numbers in millions.

bacterial numbers, which suggests a succession of different populations.

All observations of increased numbers of bacteria in soil following biocidal treatments are based on counts of bacteria by methods involving their culture in or on nutrient media, usually by a plate-count method (see, for example, reference 27). There are three reports of direct counting methods being used. Meiklejohn[82] used both plate counts and the direct method of Jones and Mollison[65] to determine the numbers of bacteria plus actinomycetes in soil from sites in Kenya where the vegetation had been cleared by burning. Both methods showed a decrease in numbers immediately after burning; but when adjacent burned and unburned sites were sampled 2, 3 and 4 months later, the numbers of bacteria plus actinomycetes as measured by plate counts were greater in the burned areas. The direct method, however, gave lower numbers in the burned soils. Shields, Paul and Lowe[107] measured bacterial numbers in soil that had been fumigated with chloroform, using both plate counts and the direct method of Babiuk and Paul.[3] After fumigation the soil was inoculated with untreated soil and incubated for 21 days. Bacterial numbers measured by plate counts followed the usual pattern, i.e. an immediate decrease followed by a rapid increase in numbers. The numbers then decreased to pre-fumigation levels within a few days. Bacterial numbers measured by the direct method decreased immediately after fumigation and then recovered to a level indistinguishable from that before fumigation. Similarly, Jenkinson, Powlson and Wedderburn,[64] using a modification of the direct method of Jones and Mollison,[65] found no evidence of increased bacterial numbers in fumigated soil. Direct methods of counting soil organisms usually give counts that are 10 to 100 times greater than those given by plate counts.[27, 111] Thus, plate counts reveal only a small section of the soil microbial population. It seems that the increase in numbers of bacteria following a biocidal treatment is confined to this section and is not a general trend of the entire soil population. The organisms that utilise substrates released by biocidal treatments (see Section 11.7) may well be those that develop most readily on nutrient agar. The increase in numbers of bacteria, as measured by plate counts, is of the order of 10^8 organisms per gram of soil (see *Table 11.1* and *Figure 11.1*). Such an increase would probably not be detectable among the 10×10^8 to 200×10^8 organisms counted in soils by direct methods.[64, 65, 107, 111]

Fungi are more readily killed by heat or fumigants than many bacteria (see reference 77). In Reber's work[93] (*Figure 11.1*) fumigation with chloropicrin eliminated the soil fungal population and recovery was not appreciable after 16 weeks' incubation, even with an inoculum of untreated soil. Methyl bromide and methyl isothiocyanate had similar effects and greatly decreased fungal numbers for at least 3 months in field experiments. Long-term decreases in fungal numbers following heat treatment or fumigation have often been reported. Warcup[125] found fungal numbers still depressed in soils 18 months after exposure to steam or formalin. Similar results from steam and formalin were found by Mollison;[84] even after 25 months, fungal numbers had not completely recovered. Martin, Baines and Ervin[78] reported that the fungal populations of many soils were still depressed 2 or 3 years after fumigation with D-D. In contrast, there are also reports of increased numbers of fungi in fumigated or heat-treated soil.[66, 122] Martin[75] found that the time of re-establishment of the fungal population, and the total numbers attained, could vary considerably between replicates in field and greenhouse experiments; this may explain some of the discrepancies between the results of different workers.

As with bacteria, plate counts of fungi in soil should be treated with considerable caution. Gray and Williams[51] state that most counts of fungi obtained in this way are in fact counts of fungal spores rather than pieces of vegetative hypha. Many spores can develop from a relatively small length of hypha and each spore can form a colony on an agar plate, thus giving a much higher count than would have been obtained in the absence of sporulation.[1] Waksman and Starkey[124] recognised this and pointed out that the increased fungal numbers they had observed with plate counts after certain biocidal treatments[122] may have reflected increased numbers of spores in the treated soils rather than more hyphal growth. The fungal population recolonising fumigated or heat-treated soil usually contains fewer species than in untreated soil.[75, 122, 125] Often one species becomes dominant, and sometimes this species remains dominant for a long period or sometimes several different species become dominant in succession at different times after the treatment.[75] Many workers have found *Trichoderma* and *Penicillium* spp. to be dominant in fumigated or heat-treated soils, most often *Trichoderma viride*.[77, 84, 94, 103, 125] Saksena[103] studied the resistance of various soil fungi to fumigants, and their ability to recolonise steamed and fumigated soil. He concluded that the success of *T. viride* was due to its relatively high growth rate rather than to its tolerance of biocidal chemicals, which was much less than that of many other soil fungi. Although *T. viride* is most commonly reported as dominant after a biocidal treatment, many other species also become dominant under certain circumstances (see references 75 and 125). The fungal species that becomes dominant after a biocidal treatment will be determined by a combination of many factors, such as the type of treatment, the chemical and physical properties of the soil at the time of treatment or after, the relative abundance of different species in the original soil, and whether the soil is inoculated either by chance or deliberately, after the treatment. In view of this, therefore, it is not surprising that the outcome is unpredictable. This is illustrated by the results of Evans,[43] who studied the fungal recolonisation of steamed soil contained in a glass tube and inoculated at one end with untreated soil. If the soil was loosely packed, *T. viride* was almost always the leading coloniser; but if closely packed, *Pythium, Zygorrhynchus* and *Mucor* spp. colonised first. Evans also found that the sequence of fungal invasion in a given soil varied from one experiment to another done at a different time and from one replicate tube to another within an experiment.

Reference to *Figure 11.1,* and to the work referred to above, suggests that responses of the bacterial and fungal populations to biocidal treatments are very different. However, this conclusion may be an artifact of the methods used for counting the organisms in soil. The increase in bacterial numbers that plate counts reveal shortly after a biocidal treatment may only reflect the activity of a small section of the population. Also, certain species of fungi sometimes increase in number after such treatments. Thus, with both bacteria and fungi there is the possibility of one species, or a small number of species, increasing in number relative to the rest of the population following the initial decrease in numbers.

The effects of fumigants and heat treatments on other soil organisms are less well documented than the effects on bacteria and fungi. *Figure 11.1* shows that fumigation greatly decreased the number of actinomycetes, as with the fungi. There are also reports of slight decreases;[10, 25, 66] increases, though less than with bacteria;[117] and variable effects.[122]

Singh and Crump[109] found that steamed soil contained more protozoa than untreated soil over a period of 7 months, whereas soil treated with formalin contained significantly fewer protozoa than untreated soil for at least a year. Stout[116] found that although samples of steamed soil taken 100 days after steaming contained more protozoa than the corresponding unsteamed soil, the steamed soil contained far fewer species.

There is much information on the effects of soil fumigants on numbers of plant parasitic nematodes. In recent field experiments[130, 131] the fumigation of soil into which wheat was subsequently sown decreased invasion by cereal cyst nematode, *Heterodera avenae* Woll., but led to increased numbers of eggs in the soil after harvest. This increased population adversely affected the succeeding crop. The healthier plants in the fumigated plots had larger root systems than the plants in the unfumigated plots and were therefore better hosts for nematodes, which resulted in an increase in *H. avenae* numbers. However, increases in plant growth caused by extra nitrogen fertiliser were not accompanied by such large increases in *H. avenae* numbers, which suggested that fumigation may have affected the *H. avenae* population through other factors also.

Edwards and Lofty[35] found that fumigation with D-D or Vapam had drastic effects on the soil fauna; usually over 99% of the original numbers were killed, and it took more than 2 years for the population to recover completely. In another experiment[36] Dazomet caused a 60% decrease in numbers of soil invertebrates during the following 5 months. D-D, Metham sodium and methyl bromide killed nearly all the earthworms in experimental plots[38] but, surprisingly, formaldehyde and Thiram had no effect on earthworm numbers.[37]

Irradiation

Soil organisms vary considerably in their sensitivity to irradiation, depending on their size and complexity.[79] Cawse[16] states that few invertebrates would survive a dose of 0.2 Mrad; 0.2–0.5 Mrad would inactivate the majority of fungi; and 1–2 Mrad would inactivate practically all soil bacteria. Popenoe and Eno[90] give data on the survival of bacteria, fungi, algae and nematodes after being subjected to doses of gamma-irradiation ranging from 0.001 to 2.048 Mrad. However, Cawse[16] points out that variations in the number of organisms from one soil to another and differences in radiosensitivity make generalisations unwise. Robinson, Corke and Jones[98] found that soil 'sterilised' with 5 Mrad gamma-irradiation still contained viable bacteriophages.

Enzymes are relatively radioresistant and there is abundant evidence of enzyme activity in radiation-sterilised soil. Some examples are: urease activity,[80, 81, 95] phosphatase activity,[80] nitrification,[14, 15, 24] nitrate reduction[18, 19] and respiration.[21, 88, 96] At least part of this activity is due to enzymes situated within cells that have been rendered incapable of proliferation because of irradiation damage to their protein synthesis system; the nucleic acids are probably the most vulnerable part of this system (see reference 79). McLaren and co-workers[80, 81] suggested that irradiation could increase the permeability of cell walls, so that substrates could diffuse to enzymes more easily. Extracellular enzymes may also contribute to the enzyme activity of irradiated soil.[79, 80]

An increase in bacterial numbers during the post-irradiation period, similar to that in fumigated soil, has been observed in irradiation-sterilised soil exposed to chance inoculation, but fewer species were present than in untreated soil.[79, 80]

Air-drying

Air-drying is a treatment to which soils are frequently subjected, both in the field and for storage in the laboratory. Consequently, the effects of this treatment are of considerable importance. Although much less biocidal than fumigation, heating or irradiation, air-drying decreases bacterial numbers,[112, 114, 121] but, after remoistening, the numbers (as measured by plate counts) increase and temporarily exceed the numbers in untreated soil, as with fumigated, irradiated or heated soil.[114, 121] The proportion of

201

the soil organisms killed by air-drying depends on the extent to which the soil is dried and on the conditions of drying.

Unfortunately, the effects of air-drying on soil metabolism are often ignored; some workers have used air-dried soil when investigating the effects of other biocidal treatments, and some have air-dried soil in order to drive off fumigants. Consequently, the effects of the treatment under investigation are partially masked by the effects of air-drying. If the use of air-dried soil is unavoidable, the results are more meaningful if the soil is remoistened and incubated for one or two weeks, to allow the flush of bacterial activity to subside, before the soil is subjected to other treatments.

Freezing and thawing

Freezing at −22°C for 24 h caused a decrease in the number of viable micro-organisms in soil but was less detrimental than air-drying.[112] Freezing at −196°C for 10 min decreased the number of viable fungi.[72] Grossbard and Hall[52] found that actinomycete counts, and to some extent bacterial counts, tended to decrease with time of storage at −15°C. Recently Biederbeck and Campbell[5] showed the detrimental effects of fluctuating temperature occurring in the field. Bacterial numbers (as determined by plate counts) were decreased by 54% when fresh soil was frozen by successive placement in cold rooms at 4, 0 and −12°C for 1 day each, −23°C for 6 days, and then −12 and 0°C for 1 day each. Thawing for 4 days alternately at 14 and 3°C decreased the number of bacteria by 92%, of fungi by 55% and of actinomycetes by 33%, compared with the untreated soil. The fluctuating temperature also caused qualitative changes in the soil microflora.

Mechanical disturbance

It is to be expected that some larger soil organisms will be killed by mechanical disturbance such as ploughing, rotovating or grinding, but there is little information on this. In field experiments Edwards and Lofty[35] found that ploughing an old pasture decreased the number of arthropods. However, this was almost certainly due to factors such as the redistribution of partially decomposed organic matter as well as a direct result of killing during ploughing.

There is little information on the *direct* effects of mechanical disturbance on numbers of soil micro-organisms—numbers can be affected *indirectly* by the resulting changes in soil aeration and water-holding characteristics. Skinner, Jones and Mollison[111] found that plate counts of bacteria obtained after gently grinding soil with water were the same as those obtained after shaking soil with water.

11.4 Changes in soil respiration

Respiration is a frequently used index of biological activity in soil, as it reflects the activities of all sections of the soil population. Respiration measurements (consumption of oxygen or evolution of carbon dioxide) are, therefore, extremely valuable for studying the effects of biocidal treatments on the activity of the soil population as a whole and on the rate of decomposition of soil organic matter.

Biocidal treatments alter parts of the soil organic matter, making it more decomposable (see Section 11.7). After the treatment, the surviving soil organisms or those introduced in an inoculum can utilise the newly available substrates, and this activity is reflected in increased soil respiration—*Figure 11.2* shows some examples of this. If few organisms survive the treatment, then the period of rapid respiration will be preceded by a lag period, as with methyl bromide fumigation (*Figure 11.2*). With sterilising treatments, where there are no survivors, there is no increase in respiration until after an inoculum is added. Thus, as with studies on microbial numbers (Section 11.3), it is important to specify whether the treated soil has been inoculated. The effects of various biocidal treatments on soil respiration are considered below.

Fumigation

After fumigation, respiration is inhibited, for times ranging from a few hours to a few weeks, because of the biocidal effects of the fumigant. The period of inhibition depends on the number of organisms surviving the treatment and on the presence or absence of an inoculum. After the initial decrease, respiration rises to high values and exceeds that of untreated soil (see references 10, 32 and 54 and *Figure 11.2*). These effects can be modified if traces of the fumigant, or its breakdown products, remain in the soil. The continued presence of a biocidal chemical in the soil can inhibit the

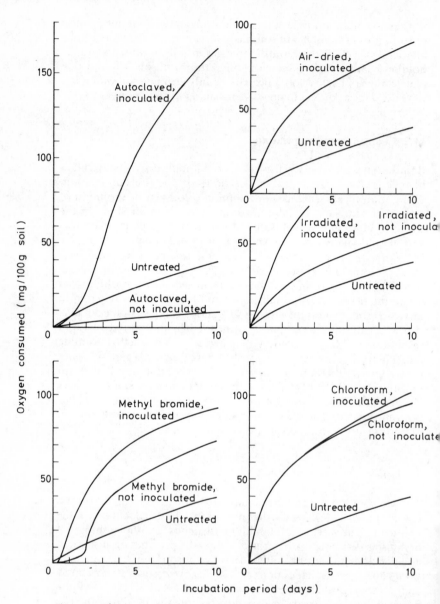

Figure 11.2 The oxygen consumption of an arable soil after various biocidal treatments, either with or without an inoculum of untreated soil (from Powlson and Jenkinson[92])

recolonising population and increase the lag period. Some chemicals can be metabolised by soil organisms, so causing increased respiration; both effects were observed in soil treated with formaldehyde solution that contained methanol as a stabiliser.[62]

Heating and drying

Both oven-drying and air-drying, followed by remoistening, increase respiration during subsequent incubation.[6, 8, 67, 121-123] Soil that was air-dried for 24 h at 25°C, and then remoistened, showed no lag in oxygen consumption and the maximum respiration rate occurred at the beginning of the incubation (*Figure 11.2*); this was also found by Chase and Gray.[26] However, there was a lag phase when soil was dried at 80°C,[44] this treatment being much more biocidal. Stevenson[114] measured both oxygen uptake and bacterial numbers in air-dried and re-wetted soil, and found the maximum respiration rate just before the rapid increase in bacterial numbers. Autoclaving sterilises soil and greatly decreases respiration, although some oxygen and carbon dioxide exchange continues owing to chemical rather than enzymic processes.[13, 88, 96] *Figure 11.2* shows that when autoclaved soil is inoculated with untreated soil, respiration is increased even more than in fumigated or irradiated soil (see references 60 and 89), but this high rate of respiration is preceded by a marked lag phase (see also reference 96).

Irradiation

Gamma-irradiation, in the absence of an inoculum of untreated soil, may either increase or decrease respiration, depending on the dose and the soil. Roberge[96] observed decreased respiration in soil and litter given 1.1 Mrad gamma-radiation, whereas Cawse and Mableson[21] found that all doses between 0.1 and 10 Mrad increased respiration, 2 Mrad causing the largest increase. *Figure 11.2* shows that 2.5 Mrad, without inoculation, increased respiration. Respiration in uninoculated, radiation-sterilised soil is due to the activity of radiation-resistant enzymes (see p. 201), although at higher doses enzymes become inactivated and CO_2 evolution due to radiolytic decarboxylation of soil organic matter becomes more important. Cawse and Mableson[21] measured CO_2 evolution from gamma-irradiated soil, both during irradiation and during the following 24 h, and concluded that with a dose of up to 2 Mrad the CO_2 evolved was mostly due to enzyme activity, but at 10 Mrad 45% of the CO_2 could be formed by decarboxylation.

Figure 11.2 shows that when irradiated soil is inoculated with untreated soil, there follows a period of increased respiration similar to that in fumigated, heated or dried and remoistened soil (see also references 60, 62 and 91). Because of enzyme activity there is no lag in respiration in the period before microbial numbers have built up, in contrast to fumigated soil, where enzyme activity is inhibited immediately after treatment.[63, 92]

Freezing and thawing

Freezing and thawing soil before incubation increases respiration compared with soil kept at a constant temperature above zero.[57, 72, 99] The effects are smaller and more short-lived than with more biocidal treatments. Ross[99] found that the respiration rate of soil that had received a freezing treatment (18 h at −20°C) subsided to that of unfrozen soil within 24 h. The 'untreated' soil in *Figure 11.2* had been stored at −15°C for about 1 year. Respiration was enhanced during the first few days of incubation at 25°C, and gradually decreased throughout the 10 day period. Jenkinson and Powlson[63] found that a progressive change occurred during frozen storage. The longer soil was kept at −15°C, the greater the respiration during the first few days of incubation at 25°C. However, all these effects of freezing and thawing are smaller than those of air-drying, so that for most work on soil metabolism, freezing is preferable to air-drying as a method of storing soil samples.

Mechanical disturbance

Rovira and Greacen[101] subjected moist soil to compression and shearing, which caused a short-term increase in respiration. They concluded that this was because the treatment exposed organic matter that was not previously accessible to microbial decomposition, perhaps because it was contained in small pores. Craswell and Waring[29] showed that grinding increased soil respiration in the four soils they studied. The increases were fairly short-lived, mostly lasting for less than 5 days.

11.5 Changes in the mineralisation of soil nitrogen

Biocidal treatments can alter the mineral nitrogen content of soil, both during the treatment and during subsequent incubation.

Changes occurring during treatment

Increases in ammonium-N can occur during soil treatment—for example, during autoclaving,[104, 108] heating,[60] steaming,[31, 41] fumigation.[41, 60, 63, 124] air-drying and irradiation.[12, 14, 15, 41, 63, 104, 110] Eno and Popenoe[41] found that this release of ammonium-N was much greater during steaming than irradiation (3 Mrad) or fumigation with methyl bromide. Singh and Kanehiro[110] found that the amount of ammonium-N liberated increased with increasing dose of gamma-irradiation over the range 0.5–5.0 Mrad. Fumigation and heating can cause a small decrease in nitrate[58, 60, 63] but the mechanism is not known.

Changes occurring during subsequent incubation

The accelerated decomposition of soil organic matter caused by biocidal treatments (followed by inoculation with untreated soil where necessary), which causes increased soil respiration, also leads to increased mineralisation of nitrogen during subsequent incubation. In general, the organic matter mineralised after a biocidal treatment has a lower C/N ratio than that mineralised in untreated soil, which indicates that the fraction of the soil organic matter rendered decomposable is richer in nitrogen. Thus, the increase in mineralisation of nitrogen compared with untreated soil is even more marked than the increase in CO_2 evolution. This increased mineralisation of nitrogen has been repeatedly observed in laboratory, greenhouse and field experiments (see references 2, 31, 34, 39, 41, 48, 55, 58, 60, 67, 86, 120–124 and 131). For examples of a flush of mineral nitrogen caused by air-drying and re-wetting see references 6, 9, 47, 86. The magnitude of the flush caused by air-drying increases with the period of air-dry storage.[47] The organic matter mineralised after air-drying and re-wetting has a greater C/N ratio than that mineralised after the more biocidal treatments.[92] Bowen and Cawse,[12] Cawse[15] and Eno and Popenoe[41, 42] found that gamma-irradiation increased the mineralisation of soil nitrogen during subsequent incubation. Freezing and thawing can also increase mineralisation of nitrogen, but the effect is less than with other treatments.[46, 72] Alternate freezing and thawing increased the mineralisation of nitrogen during subsequent incubation more than freezing alone.[46] Craswell and Waring[28] found that grinding to <0.18 mm and <0.05 mm increased the subsequent mineralisation of nitrogen in 7 soils out of 10.

The extra nitrogen mineralised during the flush of decomposition can enhance crop growth, when nitrogen is limiting. This has long been recognised: see references 2, 30, 39 and 54. Ennik, Kort and van Doorn[40] reported that the increased yield of grass caused by fumigation was, at least partly, due to an increase in available nitrogen. It is sometimes possible to demonstrate increases in plant growth and nitrogen uptake following a biocidal treatment which are due entirely, or almost entirely, to extra mineral nitrogen in the soil.[11] Normally, however, a plant response to a soil biocidal treatment will be the combined result of both pathogen control and extra nitrogen. If adequate nitrogen fertiliser is applied, the effect of the nitrogen released by the biocidal treatment may not be apparent in increased plant growth.

Air-drying and oven-drying, followed by re-wetting, caused increased mineralisation of soil nitrogen and increased the yield and nitrogen content of millet.[9] In tropical areas with distinct wet and dry seasons some crops are planted towards the end of the dry season to utilise the flush of mineral nitrogen that occurs with the onset of the rains; see references 8 and 106 for examples of this in East Africa.

One of the best-known effects of highly biocidal treatments, such as fumigation, heating and irradiation (at high doses), is the inhibition of nitrification. This is referred to in all the reviews cited at the beginning of this section and also those by Gainey,[45] Stark, Smith and Howard,[113] Domsch,[32] Hall and Clegg,[54] Martin[77] and Warcup.[126] Bacteria that oxidise ammonium to nitrite and nitrite to nitrate are relatively sensitive to the more severe biocidal treatments. Ammonification is carried out by a greater variety of organisms, some of which can either survive the treatments or recolonise more quickly than nitrifiers. Nitrification may be stopped, or greatly decreased, for considerable periods. Martin[77] showed that fumigation inhibited nitrification for several weeks to several months. Malowany and Newton[74] found that steaming stopped nitrification for at least 40 weeks if the soil was not inoculated with fresh soil. Even in inoculated soils it was 6–10 weeks before nitrification started again. Davies and Owen[31] found that nitrification was suppressed for rather shorter periods in steamed soils. However, their soils were exposed to either aerial contamination or disturbance due to sampling. They found ammonium was unoxidised for at least 15 weeks in autoclaved soil when incubated undisturbed. Jenkinson and Powlson[63] found that soil fumigated with chloroform and inoculated with fresh soil did not nitrify for at least 30 days. It is not known why nitrification takes so long to recover in heated or fumigated soil, even after re-inoculation. Possibly some constituent of the organic matter rendered soluble by these treatments (see Section 11.6) is toxic to nitrifiers.

As a result of inhibition of nitrifiers, the nitrogen mineralised in a soil previously given a strongly biocidal treatment remains as ammonium, as does any ammonium fertiliser added. This alteration of the balance between ammonium and nitrate in the soil may have a physiological effect on some plants (see reference 77). Nitrate can be lost from the rooting zone of plants by leaching, but ammonium is not subject to leaching. Thus, not only is more nitrogen mineralised in a fumigated, heated or irradiated soil than in untreated soil, but also much of it is in a form less liable to losses, and so a higher recovery of nitrogen by plants can be expected. These effects are illustrated in field experiments by Draycott and Last.[34]

Air-drying followed by re-wetting does not inhibit nitrification (see, e.g., references 8 and 47), so nitrogen mineralised after air-drying and re-wetting is oxidised to nitrate as in untreated soil. However, Cawse and Sheldon[22] found that in organic calcareous soils nitrate reduction could occur concurrently with the increased nitrate formation following air-drying and re-wetting, even though the water content of the soil was well below saturation. Nitrite accumulated for 1–2 days after rewetting but decreased rapidly on further incubation; some nitrous oxide was formed. They concluded that oxidation of organic matter released during air-drying (see Section 11.6) caused an oxygen stress in the soil, so that nitrate reduction could occur at some sites, even though the soil as a whole was aerobic. Any decrease in pore space caused by air-drying and re-wetting will increase the number of such sites.

McLaren[79] reviewed the effects of irradiation on nitrification; these effects vary with the dose applied and with the soil. Cawse[14] found that low doses (between 0.002 and 0.2 Mrad) increased the amount of nitrate formed in soils during 7 days' incubation. In 8 soils out of 10, irradiated over the dose range 0.25–2.5 Mrad, nitrate accumulation was greater than in controls, with maximum accumulation at doses between 0.25 and 0.75 Mrad.[23] These increases in nitrate formation were probably due to enzyme activity in non-proliferating nitrifying bacteria rather than to proliferation of the surviving bacteria.[14, 15] Jenkinson and Powlson[63] found that a soil given 1 Mrad produced more nitrate than untreated soil during the subsequent 10 days' incubation, but with 2.5 Mrad nitrate formation was about equal to that in untreated soil. However, even at 1 Mrad there was an accumulation of ammonium, suggesting that nitrification in irradiated soil is performed by a population unable to proliferate.

Accumulation of nitrite has been reported in irradiated soils, especially in the dose range 0.5–0.8.[20, 23] In soils containing more than 1% carbonate–carbon, and with a pH greater than 7, this accumulation could be large (over 100 ppm in one soil) and could persist for several days. In other soils accumulation was less, lasting only a few hours, and in an acid soil (pH 4.0) nitrite was not detected at all. This nitrite was formed by reduction of nitrate, probably by the enzyme systems of non-proliferating denitrifying bacteria.[18, 19]

11.6 Changes in water-soluble organic matter

Biocidal treatments increase the amount of water-soluble soil organic matter.[2, 53, 67, 79, 121, 125] In general, heat treatments release more organic matter than irradiation or fumigation, which in turn releases more than air-drying or freezing. The amount of organic matter solubilised by irradiation increases with increasing dose.[12, 14] Birch[6, 7] found that organic matter rendered water-soluble by air-drying or oven-drying was very decomposable and considered that the decomposition of this material was in part responsible for the increased respiration following air-drying and re-wetting. This is supported by Jagnow,[59] who measured the increase in respiration caused by air-drying and re-wetting in 50 East African soils; respiration was positively correlated with the amount of hot-water-extractable organic matter in the soil at the beginning of the incubation.

Powlson and Jenkinson[92] found that the amounts of organic matter solubilised by air-drying, fumigation ($CHCl_3$ or CH_3Br), gamma-irradiation (at 2.5 Mrad) or autoclaving were highly correlated with both respiration and mineralisation of nitrogen during subsequent incubation. However, the extra organic matter decomposed following treatment and inoculation did not come entirely from this water-soluble fraction; some insoluble organic matter decomposed and some of the organic matter solubilised was resistant to decomposition. The amounts of resistant organic matter solubilised by different treatments were in the order: fumigation ($CHCl_3$ or CH_3Br) < irradiation (2.5 Mrad) < autoclaving. Changes in concentrations of soluble amino acids and sugars in soils during frozen storage were observed by Paul and Tu[86] and Ivarson and Sowden.[57]

Immediately after steaming or autoclaving, soil is toxic to plant growth[97] and steamed soil is usually either flooded or left for some time before planting.[102] Rovira and Bowen,[100] using autoclaved soil, showed that the toxicity could be eliminated by certain fungi and bacteria found in inoculated autoclaved soil, or by leaching with water. They concluded that the toxicity was due to water-soluble organic material formed during heating. Peterson[89] found a considerable lag in growth and oxygen consumption by soil organisms inoculated into autoclaved soil; the lag was longer if the soil was autoclaved for 2 instead of 1 h. Salonius, Robinson and Chase[104] measured the respiration of soil organisms inoculated into an artificial soil (a mixture of fine sand and kaolinite) to which they added an aqueous extract from either air-dried, irradiated (3 Mrad) or autoclaved soil. Decomposition of extracts from air-dried or irradiated soil started immediately but of extracts from autoclaved soil only after a lag of 1–2 days.

11.7 Explanations of the changes in soil metabolism caused by biocidal treatments

For a short period following a biocidal treatment (and, where necessary, inoculation with soil organisms) soil organic matter decomposes faster than in untreated soil. This accelerated decomposition is associated with increased numbers of certain soil organisms, particularly bacteria, and leads to increases in consumption of oxygen and mineralisation of carbon, nitrogen and other elements. The extra decomposition in treated soil has been called a flush of decomposition,[60] defined as the amount of decomposition in the treated soil (measured by O_2 consumption, CO_2 evolution or nitrogen mineralisation) *less* that in the untreated soil incubated for the same time under the same conditions.

Jenkinson[60] has summarised the theories proposed to explain this phenomenon and placed them into three groups.

Effects of Biocidal Treatments on Soil Organisms

Group I. Microbial activity is inhibited in unsterilised soil

Biocidal treatments may disrupt the inhibiting process, so that a flush of decomposition can occur, ending as the inhibiting process builds up again. Examples of such theories are:

(1) Biocidal treatments destroy a toxin-restraining microbial development. The checked population proliferates until a new build-up of toxin again restricts growth.

(2) Biocidal treatments increase the 'physiological vigour' of the surviving organisms. This theory cannot explain the occurrence of a flush in sterilised soil where organisms are introduced in an inoculum.

(3) Ecological theories. Decomposition is retarded by antagonism between different sections of the population, one section keeping another in check; after a biocidal treatment the previously checked section of the population grows freely for a time, causing a flush of decomposition. One section of the population could keep another in check either directly by a host–predator relationship, or indirectly, e.g. by producing an antibiotic.

Jenkinson[60] points out a general objection to all these theories. If microbial activity is inhibited, some otherwise available substrate must be protected from attack. Such a substrate is unlikely to persist for long in a competitive environment such as soil, although it could persist for a short time following the addition of a more readily decomposable substrate such as fresh plant material. A population immune to the inhibiting process, and able to attack the protected substrate, would evolve.

Group II. Otherwise available substrates are physically protected from microbial attack in unsterilised soil

Biocidal treatments remove this protection, allowing attack to proceed until the newly exposed substrate is consumed. For example:

(1) Lipid solvents such as chloroform remove a waxy film covering an otherwise available substrate. This theory cannot explain the effect of irradiation.

(2) Drying causes cracking of organic colloids, exposing a larger surface area to microbial attack.

(3) Available substrate is covered by a film of organisms that offer a barrier to others only so long as they are living.

Group III. Biocidal treatments produce available substrates from otherwise unavailable ones

These substrates, which are rapidly attacked, could be released by the death and lysis of organisms or by increasing the decomposability of non-living sections of the soil organic matter.

To test these theories and to identify the source of the substrate decomposed during the flush, Jenkinson[60] used soil that had been incubated in the field for 1 year with ^{14}C-labelled plant material. The soil population decomposing the plant material, and the resulting metabolites, would be heavily labelled, and parts of the soil organic matter that are older and biologically less active would be lightly labelled. Samples of this soil were subjected to air-drying, irradiation (0.25 Mrad), methyl bromide fumigation, chloroform fumigation, oven-drying (80 and 100°C) and autoclaving (120°C). The main conclusions were:

(1) Different treatments, whose only common effect was the killing of organisms, all caused the evolution of comparable amounts of CO_2, containing about the same percentage of $^{14}CO_2$, from non-uniformly labelled soil. The CO_2 evolved after treatment of a particular soil contained about ten times as much ^{14}C as did the soil organic matter as a whole. Thus, either the organisms are themselves heavily labelled and decomposed during the flush, or the soil contains a heavily labelled potential substrate that is released in similar amounts by fumigation, irradiation (0.25 Mrad) or oven-drying at 80°C.

(2) The amount of labelled CO_2 mineralised by labelled fumigated soil is not decreased by admixture with control soil. This is strong evidence against Group I theories that postulate a flush of decomposition due to disruption of a biological protection mechanism. If such a theory were correct, the protecting agent present in the unfumigated soil would decrease the amount of decomposition occurring in the fumigated soil.

(3) Organisms added to the soil decompose more quickly after fumigation.

(4) Concurrent with the mineralisation of the heavily labelled fraction, there is a 'background' production of lightly labelled CO_2.

(5) In addition to their effects on the heavily labelled fraction, irradiation and treatments involving heat render certain lightly labelled fractions of the soil organic matter decomposable.

From these results Jenkinson advanced the following hypotheses:

(a) Treatments causing partial or complete sterilisation increase the rate of mineralisation of biomass C by killing or damaging organisms. Dead and damaged organisms mineralise more quickly than undamaged ones.

(b) With some exceptions (see paragraph (c)), treatments causing partial sterilisation do not increase the rate of attack on non-biomass C.

(c) Certain treatments (irradiation and heat treatments) make parts of the soil organic matter decomposable by processes other than killing of organisms. Both biomass C and non-biomass C may be altered in this way.

These hypotheses are consistent with the low C/N ratio of the organic matter decomposed after a biocidal treatment (see p. 207)—organisms have a lower C/N ratio than the soil organic matter. It has long been recognised that the decomposition of killed organisms contributes to the flush of decomposition (see reference 115). However, emphasis has often been placed on other factors, especially that decreased microbial competition in treated soil allows of better growth of the surviving or recolonising organisms.[76, 77, 121] Result (2) above indicates that reduced competition has little effect on the over-all amount of organic matter decomposed in treated soil. Decreased competition may be a factor determining the relative numbers of the various groups of organisms in the recolonising population, but not the over-all amount of decomposition that occurs.

Shields *et al.*[107] considered that fumigation with chloroform, in addition to killing organisms, also rendered decomposable considerable amounts of microbial metabolites. Jenkinson[60] found that fumigation, irradiation (0.25 Mrad) and oven-drying at 80°C all caused about the same amount of $^{14}CO_2$ to be evolved during subsequent incubation. It is difficult to imagine a mechanism by which such diverse treatments all increase the decomposability of microbial metabolites, or any other non-biomass organic matter, to about the same extent. If the metabolite theory is correct, then fumigation of a soil in which all organisms have already been killed by irradiation should produce a larger flush than irradiation alone. Jenkinson and Powlson[63] tested this idea and found that irradiation followed by fumigation with chloroform *did* cause a slightly larger flush than irradiation alone but that this increase was due to other factors; they concluded that the results did not support the metabolite theory.

Jenkinson[60] found that labelled soil evolved more labelled and unlabelled CO_2 after heat treatments than after fumigation, although the percentage of ^{14}C in the carbon dioxide was slightly less. As the severity of the heat treatment increased in the sequence oven-drying at 80°C, oven-drying at 100°C, autoclaving at 120°C, little more $^{14}CO_2$ was evolved during subsequent incubation but the evolution of unlabelled CO_2 increased markedly. This suggests that, besides killing organisms, heat rendered some other less heavily labelled part of the soil organic matter decomposable. Furthermore, a second oven-drying given immediately after the first significantly increased the amount of unlabelled CO_2 evolved during subsequent incubation, but little increase in the evolution of $^{14}CO_2$. In contrast, a second fumigation with chloroform did not increase the amount of labelled or unlabelled CO_2 evolved. Thus, heat treatments render an appreciable amount of non-biomass organic matter decomposable, the amount increasing with the severity of treatment.

In a range of soils, irradiation of 2.5 Mrad consistently gave a slightly larger flush than fumigation with chloroform or methyl bromide, and respiration remained higher than that of untreated soil for longer (*Figure 11.2*). Jenkinson[60] found that irradiation (2 Mrad) had some effect on non-biomass organic matter; labelled soil irradiated twice evolved significantly more unlabelled CO_2 during subsequent incubation than after a single irradiation, and evolved slightly more $^{14}CO_2$. The radiation dose given during the first treatment (2 Mrad) was enough to sterilise the soil, so additional CO_2 after the second irradiation could not have arisen from a more complete kill. Irradiation probably increases the decomposability of organic matter through the formation in the soil water of peroxide radicals which subsequently react with organic matter.[15–17] This might affect both biomass and non-biomass organic matter.

Powlson and Jenkinson[92] found that the effects of air-drying differed markedly from those of fumigation, irradiation and autoclaving. Although air-drying and re-wetting increased the mineralisation of nitrogen in a range of soils, these increases were smaller than those caused by the other treatments. Furthermore, the C/N ratio of the organic matter decomposed during the flush was higher than with the other treatments, which indicated that less of this organic matter came from nitrogen-rich fractions of the soil organic matter than with the other treatments. Jenkinson[60] also found that the CO_2 evolved from labelled soil after air-drying and re-wetting was less heavily labelled than that after fumigation, irradiation or heating. Birch[8] found that the flush of decomposition obtained from successive air-drying/re-wetting/incubation cycles decreased gradually but that this decrease was very much slower than with successive chloroform fumigation/inoculation/incubation cycles.[60, 92] All these results suggest that air-drying increases the decomposability of a wider range of soil organic matter fractions than does fumigation. The flush following air-drying and re-wetting seems to be due both to decomposition of killed organisms and, to a considerable extent, to decomposition of non-biomass organic matter. The mechanism of this is not clear; cracking of organic colloids to expose a greater surface area to solution or microbial attack has been suggested.[7]

Mechanical treatments, such as grinding, could cause a flush of decomposition both by making physically protected organic matter available for microbial attack[28, 29, 101] and by killing organisms. Freezing and thawing probably kills organisms, but may also have indirect effects by altering soil structure.

11.8 Estimation of soil biomass from the size of the flush of decomposition following fumigation

If the flush of decomposition following fumigation is due to the decomposition of killed organisms, and not to that of non-biomass organic matter, then the size of the flush should be related to the size of the soil biomass. Jenkinson[60] derived the approximate relationship $B = F/k$, where B = soil biomass carbon (mg C per 100 g soil); F = flush of decomposition, i.e. CO_2-carbon evolved from fumigated soil less that evolved by unfumigated soil incubated under the same conditions (mg CO_2-carbon per 100 g soil); and k = the fraction of the biomass carbon mineralised to CO_2 during the incubation following fumigation and inoculation with unfumigated soil. The value of k will probably depend on microbial properties such as species of organism, phase of growth and nutritional status, and on soil factors such as aeration and pH. However, a value for k of 0.5 was provisionally proposed[61] for all sections of the soil biomass; i.e. for a 10 day incubation at 25°C, $B = F/0.3$.

The fumigation method for the measurement of soil biomass is much less laborious than the use of direct microscopy.[3, 64] Chemical methods have also been proposed, such as the determination of ATP;[33, 70, 71, 73] or muramic acid.[83] These methods assume that the percentage of ATP or muramic acid in organisms in the soil is the same as in the pure cultures used to calibrate the methods, so the precision of the results is uncertain. Thus, despite uncertainty about the value of k, the fumigation method is probably the most satisfactory method at present for measuring soil biomass, and for comparative measurements the exact value of k is not important.

The biomass in 8 near-neutral soils by the fumigation method (using $k = 0.5$) ranged from 20 mg C per 100 g soil (oven-dry basis) in a soil that had been arable for many years to 330 mg C per 100 g in a highly organic soil under permanent grass.[64] The percentage of the soil organic carbon in the biomass ranged from 1.2 to 3.4%, being greater in soils from wooded or grassland sites than in arable soils. These values for soil biomass, and similar values obtained by direct microscopy, are larger than the estimates of Satchell[105] and Babuik and Paul.[3] Two acid soils (pH 3.9 and 4.6) gave very small flushes when fumigated, which presumably indicated that they contained small biomasses.[92]

By use of the fumigation method it has been shown that the biomass is more sensitive to changes in cropping than is soil organic matter as a whole. For example, the organic C content of a Nigerian arable soil, previously under tropical forest, decreased by 16% during 2 years. In the same period the biomass decreased by 48%.[92] Jenkinson and Powlson[62] have also shown that it takes several years for the soil biomass to recover from a fumigation treatment in the field. They concluded that part of the biomass with a long half-life, probably composed of spores and other dormant organisms, was eliminated by fumigation and took several years to recover to pre-fumigation levels.

REFERENCES

1. ALEXANDER, M. *Introduction to Soil Microbiology*, 64–65. Wiley, New York (1961)
2. Anonymous. Partial sterilization, *Soils and Fertilizers*, **11**, 357–360 (1948)
3. BABIUK, L. A. and PAUL, E. A. The use of fluorescein isothiocyanate in the determination of the bacterial biomass of grassland soil, *Canadian Journal of Microbiology*, **16**, 57–62 (1970)
4. BENZIAN, B. Experiments on nutrition problems in forest nurseries, *Bulletin of the Forestry Commission, London*, No. 37, Volume 1. HMSO, London (1965)

5. BIEDERBECK, V. O. and CAMPBELL, C. A. Influence of simulated fall and spring conditions on the soil system, *Proceedings of the Soil Science Society of America*, **35**, 474–479 (1971)

6. BIRCH, H. F. The effect of soil drying on humus decomposition and nitrogen availability, *Plant and Soil*, **10**, 9–31 (1958)

7. BIRCH, H. F. Further observations on humus decomposition and nitrification, *Plant and Soil*, **11**, 262–286 (1959)

8. BIRCH, H. F. Nitrification in soils after different periods of dryness, *Plant and Soil*, **12**, 81–96 (1960)

9. BIRCH, H. F. and EMECHEBE, A. M. The effect of soil drying on millet, *Plant and Soil*, **24**, 333–335 (1966)

10. BOLLEN, W. B. Interactions between pesticides and soil microorganisms, *Annual Review of Microbiology*, **15**, 69–92 (1961)

11. BOWEN, H. J. M. and CAWSE, P. A. Effects of ionizing radiation on soils and subsequent crop growth, *Soil Science*, **97**, 252–259 (1964)

12. BOWEN, H. J. M. and CAWSE, P. A. Some effects of gamma radiation on the composition of the soil solution and soil organic matter, *Soil Science*, **98**, 358–361 (1964)

13. BUNT, J. S. and ROVIRA, A. D. The effect of temperature and heat treatment on soil metabolism, *Journal of Soil Science*, **6**, 129–136 (1955)

14. CAWSE, P. A. Effects of low sub-sterilizing doses of gamma radiation on carbon, nitrogen and phosphorus in fresh soils, *Journal of the Science of Food and Agriculture*, **18**, 388–391 (1967)

15. CAWSE, P. A. Effects of gamma radiation on accumulation of mineral nitrogen in fresh soils, *Journal of the Science of Food and Agriculture*, **19**, 395–398 (1968)

16. CAWSE, P. A. The use of gamma radiation in soil research, *United Kingdom Atomic Energy Authority, Research Group Report 6061*. HMSO London (1969)

17. CAWSE, P. A. The formation and decomposition of nitrite in gamma-irradiated soil, PhD thesis, University of London (1970)

18. CAWSE, P. A. and CORNFIELD, A. H. The reduction of ^{15}N labelled nitrate to nitrite by fresh soils following treatment with gamma radiation, *Soil Biology and Biochemistry*, **1**, 267–274 (1969)

19. CAWSE, P. A. and CORNFIELD, A. H. Factors affecting the formation of nitrite in gamma-irradiated soils and its relationship with denitrifying potential, *Soil Biology and Biochemistry*, **3**, 111–120 (1971)

20. CAWSE, P. A. and CRAWFORD, D. V. Accumulation of nitrite in fresh soils after gamma irradiation, *Nature*, **216**, 1142–1143 (1967)

21. CAWSE, P. A. and MABLESON, K. M. The effect of gamma-radiation on the release of carbon dioxide from fresh soil, *Communications in Soil Science and Plant Analysis*, **2**, 421–431 (1971)

22. CAWSE, P. A. and SHELDON, D. Rapid reduction of nitrate in soil re-moistened after air-drying, *Journal of Agricultural Science* (Cambridge), **78**, 405–412 (1972)

23. CAWSE, P. A. and WHITE, T. Accumulation of nitrate in fresh soils after gamma irradiation, *Journal of Agricultural Science* (Cambridge), **72**, 331–333 (1969)

24. CAWSE, P. A. and WHITE, T. Rapid changes in nitrite after gamma irradiation of fresh soils, *Journal of Agricultural Science* (Cambridge), **73**, 113–118 (1969)

25. CHANDRA, P. and BOLLEN, W. B. Effects of nabam and mylone on nitrification, soil respiration, and microbial numbers in four Oregon soils, *Soil Science*, **92**, 387–393 (1961)

26. CHASE, F. E. and GRAY, P. H. H. Application of the Warburg respirometer in studying respiratory activity in soil, *Canadian Journal of Microbiology*, **3**, 335–349 (1957)

218

27. CLARK, F. E. Agar-plate method for total microbial count, in *Methods of Soil Analysis, Part 2* (edited by C. A. Black), 1460–1466. American Society of Agronomy, Madison (1965)

28. CRASWELL, E. T. and WARING, S. A. Effect of grinding on the decomposition of soil organic Matter. I. The mineralization of organic nitrogen in relation to soil type, *Soil Biology and Biochemistry,* **4,** 427–433 (1972)

29. CRASWELL, E. T. and WARING, S. A. Effect of grinding on the decomposition of soil organic matter. II. Oxygen uptake and nitrogen mineralization in virgin and cultivated cracking clay soils, *Soil Biology and Biochemistry,* **4,** 435–442 (1972)

30. DARBISHIRE, F. V. and RUSSELL' E. J. Oxidation in soils and its relation to productiveness. Part II. The influence of partial sterilization, *Journal of Agricultural Science* (Cambridge), **2,** 305–326 (1907)

31. DAVIES, J. N. and OWEN, O. Soil sterilization. I. Ammonia and nitrate production in some glasshouse soils following steam sterilization, *Journal of the Science of Food and Agriculture,* **2,** 268–279 (1951)

32. DOMSCH, K. H. Soil Fungicides, *Annual Review of Phytopathology,* **2,** 293–320 (1964)

33. DOXTADER, K. G. Estimation of microbial biomass in soil on the basis of adenosine triphosphate measurements, *Bacteriological Proceedings, Abstracts of 69th Annual Meeting,* 14 (1969)

34. DRAYCOTT, A. P. and LAST, P. J. Some effects of partial sterilization on mineral nitrogen in a light soil, *Journal of Soil Science,* **22,** 152–157 (1971)

35. EDWARDS, C. A. and LOFTY, J. R. The influence of agricultural practice on soil microarthropod populations, in *The Soil Ecosystem* (edited by J. G. Sheals), 237–246. The Systematics Association, London, Publication No. 8 (1969)

36. EDWARDS, C. A. and LOFTY, J. R. Nematocides and the soil fauna, *Proceedings of the 6th British Insecticides and Fungicides Conference,* Volume 1, 158–166 (1971)

37. EDWARDS, C. A. and LOFTY, J. R. Pesticides and earthworms, *Report of Rothamsted Experimental Station for 1972, Part 1,* 211–212 (1973)

38. EDWARDS, C. A., LOFTY, J. R., WHITING, A. E. and JEFFS, K. A. Pesticides and earthworms, *Report of Rothamsted Experimental Station for 1970, Part 1,* 193–194 (1971)

39. EISSA, M. F. M. The effect of partial soil sterilization on plant parasitic nematodes and plant growth, *Mededelingen van de Landbouwhogeschool* (Wageningen), **71–14** (1971)

40. ENNIK, G. C., KORT, J. and VAN DOORN, A. M. Effect of seed and soil disinfectants on establishment, growth and mutual relations of white clover and grass in leys, *Agricultural Research Report 741.* Centre for Agricultural Publishing and Documentation, Wageningen (1970)

41. ENO, C. F. and POPENOE, H. The effect of gamma radiation on the availability of nitrogen and phosphorus in soil, *Proceedings of the Soil Science Society of America,* **27,** 299–301 (1963)

42. ENO, C. F. and POPENOE, H. Gamma radiation compared with steam and methyl bromide as a soil sterilizing agent, *Proceedings of the Soil Science Society of America,* **28,** 533–535 (1964)

43. EVANS, E. Survival and recolonization by fungi in soil treated with formalin or carbon disulphide, *Transactions of the British Mycological Society,* **38,** 335–346 (1955)

44. FUNKE, B. R. and HARRIS, J. O. Early respiratory responses of soil treated by heat or drying, *Plant and Soil,* **28,** 38–48 (1968)

45. GAINEY, P. L. Effect of CS_2 and toluol upon nitrification, *Zentralblatt für Bakteriologie, Parasitenkunde, und Infektionskrankheiten IIte Abteilung,* **39,** 584–595 (1914)

46. GASSER, J. K. R. Use of deep freezing in the preservation and preparation of fresh soil samples, *Nature,* **181,** 1334–1335 (1958)

47. GASSER, J. K. R. Effects of air-drying and air-dry storage on the mineralisable-nitrogen of soils, *Journal of the Science of Food and Agriculture,* **12,** 778–784 (1961)

48. GASSER, J. K. R. and PEACHEY, J. E. A note on the effects of some soil sterilants on the mineralisation and nitrification of soil-nitrogen, *Journal of the Science of Food and Agriculture,* **15,** 142–146 (1964)

49. GORING, C. A. I. Theory and principles of soil fumigation, *Advances in Pest Control Research,* **5,** 47–84 (1962)

50. GORING, C. A. I. Physical aspects of soil fumigation in relation to the action of soil fungicides, *Annual Review of Phytopathology,* **5,** 285–318 (1967)

51. GRAY, T. R. G. and WILLIAMS, S. T. Microbial productivity in soil, in *Microbes and Biological Productivity: 21st Symposium of the Society for General Microbiology,* 255–286 (1971)

52. GROSSBARD, E. and HALL, D. M. An investigation into the possible changes in the microbial population of soils stored at −15°C, *Plant and Soil,* **21,** 317–332 (1964)

53. GUSTAFSON, A. F. The effect of drying soils on the water-soluble constituents, *Soil Science,* **13,** 173–213 (1922)

54. HALL, N. M. and CLEGG, L. F. L. Microbiological aspects of the partial sterilization of soil by chemicals, *Proceedings of the Society of applied Bacteriology,* No. 2, 105–118 (1949)

55. HARMSEN, G. W. and VAN SCHREVEN, D. A. Mineralization of organic nitrogen in soil, *Advances in Agronomy,* **7,** 299–398 (1955)

56. HOESTRA, H. Replant diseases of apple in the Netherlands, *Mededelingen van de Landbouwhogeschool* (Wageningen), **68–13** (1968)

57. IVARSON, K. C. and SOWDEN, F. J. Effect of frost action and storage of soil at freezing temperatures on the free amino acids, free sugars and respiratory activity of soil, *Canadian Journal of Soil Science,* **50,** 191–198 (1970)

58. JAGER, G., VAN DE BOON, J. and RAUW, G. J. G. The influence of soil steaming on some properties of the soil and on the growth and heading of winter glasshouse lettuce. I. Changes in chemical and physical properties, *Netherlands Journal of Agricultural Science,* **17,** 143–152 (1969)

59. JAGNOW, G. Soil respiration, nitrogen mineralization and humus decomposition of East African soils after drying and remoistening, *Zeitschrift für Pflanzenernährung und Bodenkunde,* **131,** 56–66 (1972)

60. JENKINSON, D. S. Studies on the decomposition of plant material in soil. II. Partial sterilization of soil and the soil biomass, *Journal of Soil Science,* **17,** 280–302 (1966)

61. JENKINSON, D. S. The effects of biocidal treatments on metabolism in soil. IV. The decomposition of fumigated organisms in soil, *Soil Biology and Biochemistry* (in press)

62. JENKINSON, D. S. and POWLSON, D. S. Residual effects of soil fumigation on soil respiration and mineralization, *Soil Biology and Biochemistry,* **2,** 99–108 (1970)

63. JENKINSON, D. S. and POWLSON, D. S. The effects of biocidal treatments on metabolism in soil. I. Fumigation with chloroform, *Soil Biology and Biochemistry* (in the press)

64. JENKINSON, D. S., POWLSON, D. S. and WEDDERBURN, R. W. M. The effects of biocidal treatments on metabolism in soil. III. The relationship between soil biovolume, measured by optical microscopy, and the flush of decomposition caused by fumigation, *Soil Biology and Biochemistry* (in press)

65. JONES, P. C. T. and MOLLISON, J. E. A technique for the quantitative estimation of soil micro-organisms, *Journal of General Microbiology,* **2,** 54–69 (1948)

66. KATZNELSON, H. and RICHARDSON, L. T. The microflora of the rhizosphere of tomato plants in relation to soil sterilization, *Canadian Journal of Research,* **C21,** 249–255 (1943)

67. KOPELOFF, N. and COLEMAN, D. A. A review of investigations in soil protozoa and soil sterilisation, *Soil Science,* **3,** 197–269 (1917)

68. KREUTZER, W. A. Soil fungicides, *Recent Advances in Botany,* **1,** 466–472 (1961)

69. KREUTZER, W. A. The reinfestation of treated soil, in *Ecology of Soil-borne Plant Pathogens* (edited by K. F. Baker and W. C. Snyder), 495–508. University of California Press, Berkeley (1965)

70. LEE, C. C., HARRIS, R. F., WILLIAMS, J. D. H., ARMSTRONG, D. E. and SYERS, J. K. Adenosine triphosphate in lake sediments. I. Determination, *Proceedings of the Soil Science Society of America,* **35,** 82–86 (1971)

71. LEE, C. C., HARRIS, R. F., WILLIAMS, J. D. H., SYERS, J. K. and ARMSTRONG, D. E. Adenosine triphosphate in lake sediments. II. Origin and significance, *Proceedings of the Soil Science Society of America,* **35,** 86–91 (1971)

72. MACK, A. R. Biological activity and mineralization of nitrogen in three soils as induced by freezing and drying, *Canadian Journal of Soil Science,* **43,** 316–324 (1963)

73. MACLEOD, N. H., CHAPPELLE, E. W. and CRAWFORD, A. M. ATP assay of terrestrial soils: a test of an exobiological experiment, *Nature,* **223,** 267–268 (1969)

74. MALOWANY, S. N. and NEWTON, J. D. Studies on steam sterilization of soils. I. Some effects on physical, chemical and biological properties, *Canadian Journal of Research,* **C25,** 189–208 (1947)

75. MARTIN, J. P. Effects of fumigation and other soil treatments in the greenhouse on the fungus population of old citrus soil, *Soil Science,* **69,** 107–122 (1950)

76. MARTIN, J. P. Influence of pesticide residues on soil microbiological and chemical properties, *Residue Reviews,* **4,** 96–129 (1963)

77. MARTIN, J. P. Influence of pesticides on soil microbes and soil properties, in *Pesticides and their Effects on Soils and Waters,* 95–108. American Society of Agronomy Special Publication No. 8, Soil Science Society of America, Madison (1966)

78. MARTIN, J. P., BAINES, R. C. and ERVIN, J. O. Influence of soil fumigation for citrus replants on the fungus population of the soil, *Proceedings of the Soil Science Society of America,* **21,** 163–166 (1957)

79. McLAREN, A. D. Radiation as a technique in soil biology and biochemistry, *Soil Biology and Biochemistry,* **1,** 63–73 (1969)

80. McLAREN, A. D., LUSE, R. A. and SKUJINS, J. J. Sterilization of soil by irradiation and some further observations on soil enzyme activity, *Proceedings of the Soil Science Society of America,* **26,** 371–377 (1962)

81. McLAREN, A. D., RESHETKO, L. and HUBER, W. Sterilization of soil by irradiation with an electron beam, and some observations on soil enzyme activity, *Soil Science,* **83,** 497–502 (1957)

82. MEIKLEJOHN, J. The effect of bush burning on the microflora of a Kenya upland soil, *Journal of Soil Science,* **6,** 111–118

83. MILLAR, W. N. and CASIDA, L. E. Evidence for muramic acid in soil, *Canadian Journal of Microbiology,* **16,** 299–304 (1970)

84. MOLLISON, J. E. Effect of partial sterilization and acidification of soil on the fungal population, *Transactions of the British Mycological Society,* **36,** 225–228 (1953)

85. NEWHALL, A. G. Disinfestation of soil by heat, flooding and fumigation, *Botanical Reviews,* **21,** 189–250 (1955)

86. PAUL, E. A. and TU, C. M. Alteration of microbial activities, mineral nitrogen and free amino acid constituents of soils by physical treatment, *Plant and Soil,* **22,** 207–219 (1965)

87. PEACHEY, J. E. Chemical control of plant parasitic nematodes in the United Kingdom, *Chemistry and Industry,* 1736–1740 (1963)

221

Effects of Biocidal Treatments on Soil Organisms

88. PETERSON, G. H. Respiration of soil sterilised by ionizing radiations, *Soil Science*, **94**, 71–74 (1962)

89. PETERSON, G. H. Microbial activity in heat- and electron-sterilized soil seeded with microorganisms, *Canadian Journal of Microbiology*, **8**, 519–524 (1962)

90. POPENHOE, H. and ENO, C. F. The effect of gamma radiation on the microbial population of the soil, *Proceedings of the Soil Science Society of America*, **26**, 164–167 (1962)

91. POWLSON, D. S. The effects of fumigants on soil respiration and mineralization of nitrogen, PhD thesis, University of Reading (1972)

92. POWLSON, D. S. and JENKINSON, D. S. The effects of biocidal treatments on metabolism in soil. II. Gamma irradiation, autoclaving, air-drying and fumigation with chloroform or methyl bromide, *Soil Biology and Biochemistry* (in press)

93. REBER, H. Untersuchungen über die Wiederbesiedlung eines chemisch entseuchten Bodens, *Zeitschrift für Pflanzenkrankheiten, Pflanzenpathologie und Pflanzenschutz*, **74**, 427–438 (1967)

94. RICHARDSON, L. T. The persistence of thiram in soil and its relationship to the microbiological balance and damping-off control, *Candian Journal of Botany*, **32**, 335–346 (1954)

95. ROBERGE, M. R. Behaviour of urease added to unsterilized, steam sterilized and gamma radiation-sterilized black spruce humus, *Canadian Journal of Microbiology*, **16**, 865–870 (1970)

96. ROBERGE, M. R. Respiration of a black spruce humus sterilized by heat or irradiation, *Soil Science*, **111**, 124–128 (1971)

97. ROBINSON, R. R. Inhibitory plant growth factors in partially sterilized soils, *Journal of the American Society of Agronomy*, **36**, 726–739 (1944)

98. ROBINSON, J. B., CORKE, C. T. and JONES, D. A. Survival of bacteriophage in soil 'sterilized' with gamma irradiation, *Proceedings of the Soil Science Society of America*, **34**, 703–704 (1970)

99. ROSS, D. J. Effects of freezing and thawing of some grassland topsoils on oxygen uptakes and dehydrogenase activities, *Soil Biology and Biochemistry*, **4**, 115–117 (1972)

100. ROVIRA, A. D. and BOWEN, G. D. The effects of microorganisms upon plant growth. II. Detoxication of heat-sterilized soils by fungi and bacteria, *Plant and Soil*, **25**, 129–142 (1966)

101. ROVIRA, A. D. and GREACEN, E. L. The effect of aggregate disruption on the activity of microorganisms in the soil, *Australian Journal of Agricultural Research*, **8**, 659–673 (1957)

102. RUSSELL, E. W. *Soil Conditions and Plant Growth*, 9th edition, 220–223. Longmans, London (1961)

103. SAKSENA, S. B. Effect of carbon disulphide fumigation on *Trichoderma viride* and other soil fungi, *Transactions of the British Mycological Society*, **43**, 111–116 (1960)

104. SALONIUS, P. O., ROBINSON, J. B. and CHASE, F. E. A comparison of autoclaved and gamma-irradiated soils as media for microbial colonization experiments, *Plant and Soil*, **27**, 239–248 (1967)

105. SATCHELL, J. Feasibility study of an energy budget for Meathop Wood, in *Productivity of Forest Ecosystems* (edited by P. Duvigneaud), 619–630. UNESCO, Paris (1971)

106. SEMB, G. and ROBINSON, J. B. D. The natural nitrogen flush in different arable soils and climates in East Africa, *East African Agricultural and Forestry Journal*, **34**, 350–370 (1969)

222

107. SHIELDS, J. A., PAUL, E. A. and LOWE, W. E. Factors influencing the stability of labelled microbial materials in soils, *Soil Biology and Biochemistry,* **6,** 31–37 (1974)

108. SIMPSON, F. J. and NEWTON, J. D. Studies on steam sterilization of soils. II. Some factors affecting minimum sterilization requirements, *Canadian Journal of Research,* **C27,** 1–13 (1949)

109. SINGH, B. N. and CRUMP, L. M. The effect of partial sterilization by steam and formalin on the numbers of amoebae in field soil, *Journal of General Microbiology,* **8,** 421–426 (1953)

110. SINGH, B. R. and KANEHIRO, Y. Effects of gamma irradiation on the available nitrogen status of soils, *Journal of the Science of Food and Agriculture,* **21,** 61–64 (1970)

111. SKINNER, F. A., JONES, P. C. T. and MOLLISON, J. E. A comparison of a direct and a plate-counting technique for the quantitative estimation of soil micro-organisms, *Journal of General Microbiology,* **6,** 261–271 (1952)

112. SOULIDES, D. A. and ALLISON, F. E. Effect of drying and freezing soils on carbon dioxide production, available mineral nutrients, aggregation, and bacterial population, *Soil Science,* **91,** 291–298 (1961)

113. STARK, F. L., SMITH, J. B. and HOWARD, F. L. Effect of chloropicrin fumigation on nitrification and ammonification in soil, *Soil Science,* **48,** 433–442 (1939)

114. STEVENSON, I. L. Some observations on the microbial activity in remoistened air-dried soils, *Plant and Soil,* **8,** 170–182 (1956)

115. STÖRMER, K. Uber die Wirkung des Schwefelkohlenstoffs und ähnlicher Stoffe auf den Boden, *Zentralblatt für Bakteriologie, Parasitenkunde, und Infektionskrankheiten IIte Abteilung,* **20,** 282–286 (1908)

116. STOUT, J. D. The effect of partial sterilization on the protozoan fauna of a greenhouse soil, *Journal of General Microbiology,* **12,** 237–240 (1955)

117. TAM' R. K. and CLARK, H. E. Effect of chloropicrin and other soil disinfectants on the nitrogen nutrition of the pineapple plant, *Soil Science,* **56,** 245–261 (1943)

118. TAYLOR, A. L. Chemical treatment of the soil for nematode control, *Advances in Agronomy,* **3,** 243–264 (1951)

119. TIETZ, H. One centennium of soil fumigation: its first years, in *Root Diseases and Soil-borne Pathogens,* 203–207. University of California Press, Berkeley (1970)

120. TILLET, E. R. The effects of different fumigants on the mineralization of soil nitrogen, *Rhodesian Journal of Agricultural Research,* **2,** 13–16 (1964)

121. WAKSMAN, S. A. *Principles of Soil Microbiology,* 2nd edition, 716–740. Baillière, Tindall and Cox, London (1931)

122. WAKSMAN, S. A. and STARKEY, R. L. Partial sterilization of soil, microbiological activities and soil fertility: I, *Soil Science,* **16,** 137–157 (1923)

123. WAKSMAN, S. A. and STARKEY, R. L. Partial sterilization of soil, microbiological activities and soil fertility: II, *Soil Science,* **16,** 247–268 (1923)

124. WAKSMAN, S. A. and STARKEY, R. L. Partial sterilization of soil, microbiological activities and soil fertility: III, *Soil Science,* **16,** 343–357 (1923)

125. WARCUP, J. H. Effect of partial sterilization by steam or formalin on the fungus flora of an old nursery soil, *Transaction of the British Mycological Society,* **34,** 519–532 (1951)

126. WARCUP, J. H. Chemical and biological aspects of soil sterilization, *Soils and Fertilizers,* **20,** 1–5 (1957)

127. WHITEHEAD, A. G., DUNNING, R. A. and COOKE, D. A. Docking disorder and root ectoparasitic nematodes of sugar beet, *Report of Rothamsted Experimental Station for 1970, Part 2,* 219–236 (1971)

128. WHITEHEAD, A. G., TITE, D. J. and FRASER, J. E. The effect of small doses of nematicides

on migratory root-parasitic nematodes and on the growth of sugar beet and barley in sandy soils, *Annals of Applied Biology*, **65**, 361–375 (1970)

129. WILHELM, S. Chemical treatments and inoculum potential of soil, *Annual Review of Phytopathology*, **4**, 53–78 (1966)

130. WILLIAMS, T. D. The effects of formalin, nabam, irrigation and nitrogen on *Heterodera avenae* Woll., *Ophiobolus graminis* Sacc. and the growth of spring wheat, *Annals of Applied Biology*, **64**, 325–334 (1969)

131. WILLIAMS, T. D. and SALT, G. A. The effects of soil sterilants on the cereal cyst-nematode (*Heterodera avenae* Woll.), take-all (*Ophiobolus graminis* Sacc.) and yields of spring wheat and barley, *Annals of Applied Biology*, **66**, 329–338 (1970)

132. WILSON, J. D. Soil fumigation, *American Potato Journal*, **45**, 414–426 (1968)

12 Non-symbiotic Nitrogen Fixation in Soil

P. J. DART AND J. M. DAY

12.1 Introduction

The source of nitrogen for plants has taxed the minds of many over the last 150 years. Lawes, Gilbert and Pugh,[55] working at Rothamsted, realised that soil contained more nitrogen than was present in the parent rock material and that there was a large unexplained source of nitrogen available to some plants but not to others. More nitrogen was removed from the field in bean or clover crops than by wheat, yet in a rotation the former crops increased the growth and yield of subsequent wheat crops and the nitrogen content of the soils. Recovery of nitrogen added as fertiliser was also much lower for legumes. Measurements showed that these crops gained little extra nitrogen from the subsoil or from rainfall. Beautifully constructed glass growth chambers and gas control systems were used to show that plants grown in a sterilised medium did not incorporate gaseous nitrogen into plant material.

Because of the thoroughness with which Lawes and Gilbert excluded contaminating life other than that of the plant from their experimental systems, the discovery of the role of micro-organisms in nitrogen fixation eluded them. In 1837 Boussingault showed that nitrogen was gained from the air by legumes grown in open pots, and substantial amounts of nitrogen were gained over 16 years by soils containing legumes in a rotation, but such revolutionary results were not accepted by Liebig, the foremost agricultural chemist of the day, who disputed the concept that plants fix nitrogen. In 1862 Jodin reported the loss of nitrogen from the atmosphere of closed containers in which cultures of 'végétaux mycodermiques' and 'les mucédinées' were grown with a carbon source but without combined

225

nitrogen, and concluded that his cultures, presumably a mixture of fungi and bacteria, had fixed nitrogen (see reference 92).

Berthelot[9] reported, in 1885, gains of nitrogen of the order of 20–30 kg N ha^{-1} over 3 summer months by both closed and open pots containing two sandy soils or two kaolin preparations and without plants. Since sterilised, closed flasks showed no such gains, he concluded that nitrogen was fixed by living micro-organisms in the soils. This conclusion was confirmed when Winogradsky isolated from soil, using a selective low-nitrogen medium, the first pure culture of a nitrogen-fixing bacterium, the anaerobe *Clostridium pasteurianum.* Blue-green algae in sand culture were shown by Frank in 1889 to fix nitrogen slowly, and Schloesing and Laurent in 1892 found considerable gains of nitrogen by surface soil covered by algae, although the difficulty in obtaining pure cultures left some doubts about this.

The involvement of micro-organisms in the formation of nodules and nitrogen fixation by legumes was unfolded during this same period by Hellriegel and Wilfarth (1886, 1888) and by Beijerinck, who isolated the causative organism, *Rhizobium,* from several legumes in 1888. Lawes and Gilbert[54] during the next few years demonstrated the ability of nodulated clovers, peas, beans, vetches, lupins and lucerne to fix nitrogen. It was Beijerinck (1901), the great Dutch microbiologist, who showed that the aerobic bacteria *Azotobacter chroococcum* from soil and *Azotobacter agilis* from canal water also grew and fixed nitrogen in nitrogen-free media.

Since then, several other genera of bacteria and blue-green algae have been reported as nitrogen fixers.[18, 33, 83] Most of the work in the first half of this century concentrated on *Azotobacter,* describing the widespread distribution and numbers of these organisms in soil or water together with their physiology and nutrient requirements. At Rothamsted, Ashby in 1907 found *A. chroococcum* and clostridia in the Broadbalk Wilderness but not in the permanent wheat experiment on Broadbalk Field. Subsequent counts for the arable field gave low numbers of *Azotobacter* (maximum of 9400 per gram of soil) with numbers of clostridia possibly higher,[59] but both organisms are too few to add any significant amounts of nitrogen to the soil.

12.2 Methods of assessment

Claims for nitrogen fixation by micro-organisms have been based (a) on growth in nitrogen-free media, (b) on gains of N in such media, or in media containing combined nitrogen, usually in the form of plant extracts, (c) on enrichment of the N in the culture with the stable ^{15}N isotope when exposed to ^{15}N$_2$-enriched gas and (d) on reduction of acetylene to ethylene by the nitrogenase enzyme.

Growth on supposedly nitrogen-free media is of limited use. Some organisms can grow on such media if they scavenge traces of combined nitrogen from the media or the air; this has led to many mistaken claims of nitrogen fixation (see reference 18). The Kjeldahl method for measuring nitrogen is relatively insensitive, and claims of fixation based on gains of less than 5% dubious (see reference 93 for critical discussion).

The development of methods for ^{15}N analysis increased the sensitivity of assays for nitrogen fixation by up to a thousand-fold. The precision of mass spectrometers analysing the ^{15}N : ^{14}N ratios exceeds even this, but the main advantage of ^{15}N tracer methods is that detection of fixation is independent of the nitrogen initially present. Determination of the absolute amount of N$_2$ fixed in a system still relies on analysis by either the Kjeldahl method or the Dumas combustion method. Mass spectrometric analysis requires much skill in setting up vacuum lines and avoiding contaminating hydrocarbons.

More recently, optical emission spectroscopy has been used to analyse ^{15}N : ^{14}N ratios by use of the relatively large isotope shifts found in the molecular spectrum of gaseous nitrogen. This method measures smaller samples than the mass spectrometer (down to about 1 µg total N) but with less precision for enrichments lower than about 0.5% over natural abundance.[68] This limits its usefulness for field and pot experiments, where the cost of ^{15}N-enriched chemicals usually necessitates the use of low initial ^{15}N enrichments.

An exciting development for field experiments is the prospect of kilogram quantities of nitrogenous compounds at low cost, which are less rich in ^{15}N than normally. The incorporation of this nitrogen source into soils, water and plants is then determined by measuring the depletion of ^{15}N in the samples by mass spectrometry.

The level of non-symbiotic nitrogen fixation in soils has been difficult to assess. Gains of less than 100 kg N ha^{-1} year^{-1} in the short term (single season or less) in the field are difficult to measure accurately. Lawes and Gilbert[54] recognised this, noting in 1889 that for reliable results it is essential that soil samples be taken in the same way with sufficient replication, and to examine soil nitrogen contents along the profile to ensure that one is measuring accumulation of fixed nitrogen and not merely nitrogen redistributed between soil horizons. Their plan was to drive a steel box, 15 cm square and 23 cm deep, into the soil to a known depth, deriving sub-samples for analyses from this volume after careful mixing. For estimates of the accumulation of nitrogen in the classical experiments at Rothamsted the same frame is still used. Erroneous results have been obtained by analysing dry soil samples before assay and wet samples after incubation, by incomplete Kjeldahl digestion and by autoclaving control soil samples with consequent volatilisation of nitrogen (see reference 48 for critical assessment of N gains reported prior to 1940).

Because of the difficulty of exposing large areas of soil to an atmosphere enriched with $^{15}N_2$, there are very few measurements of fixation in the field by isotopic methods. The relatively low activity of soil–plant systems necessitates long assay times, and this increases the difficulties of maintaining natural conditions for factors such as temperature, humidity and CO_2 concentration in the atmosphere.

In 1966 it was found that N_2 fixation by cell-free preparations from *Clostridium pasteurianum* was inhibited by acetylene[76] and that acetylene was reduced by the enzyme to ethylene.[23] The rates of reduction of acetylene and nitrogen were comparable when considered on the basis of the number of electrons transferred:

$$N{\equiv}N \xrightarrow{\ 6e\ } 2NH_3$$

$$HC{\equiv}CH \xrightarrow{\ 2e\ } H_2C{=}CH_2$$

Both substrates appear to be reduced at the same enzyme site and to have the same reductant requirements (ferredoxin *in vivo*) and ATP. Because of the high affinity of nitrogenase for acetylene, the reduction rate is unaffected by the presence of nitrogen at air concentrations, provided saturating levels of acetylene are present. This reaction is the basis of the acetylene reduction assay now used to measure activity of nitrogen-fixing systems, the ratio of acetylene reduced to nitrogen fixed being close to the theoretical value of $3:1$.[37] Deviations from this ratio are usually due to inadequate controls.

In the assay the nitrogen-fixing system is exposed to acetylene (usually 10–20% of the atmosphere) in gas-tight containers, and the atmosphere subsequently sampled at intervals. Commercially produced acetylene always contains some ethylene, and this needs to be allowed for in the assays. The concentration of ethylene and acetylene in gas samples can be rapidly measured by gas chromatography using flame ionisation detection. The assay is about a thousand-fold more sensitive than ^{15}N isotopic methods for measuring nitrogen fixation. The equipment required, apart from the gas chromatograph, is inexpensive—plastics disposable syringes for gas handling, assay chambers which may be plastics bags, glass bottles with rubber septa in their lids or pots of soil with the plant tops enclosed in a rigid, clear plastics chamber.[19] Gas samples can be stored in the syringe or by collection in evacuated tubes or vials.

Several precautions need to be borne in mind in using the acetylene reduction method for estimating nitrogen fixation:

(1) The ammonia produced within the microbial cell is usually rapidly converted to amides and amino acids, but ethylene is not further reduced or metabolised. Ammonia incorporation requires suitable carbon skeletons, some of which can also serve for production of reductant and ATP. During acetylene reduction there is only the latter requirement. When carbohydrate supply is limited, ammonia incorporation competes with nitrogen reduction for carbohydrate, possibly reducing the rate of reduction. Over a longer term, the increase in the ammonia pool under carbon-limited conditions could repress nitrogenase synthesis.[18] Conversely, exposure to acetylene for long periods could reduce the ammonia level with consequent derepression of nitrogenase synthesis, the extra enzyme thus produced resulting in a falsely high estimate of the enzyme activity in its 'natural' state.

(2) Ethylene is a potent plant hormone.

(3) Acetylene and nitrogen have different solubilities in water and different rates of adsorption by soil, and this can lead to different rates of substrate availability.[67, 70] In practice these factors have not affected most of the systems assayed.

(4) Ethylene may be produced in soils, particularly those with much organic matter, without added acetylene.[81] This is unlikely to introduce errors in the assays for systems fixing more than 5 kg N ha^{-1} year^{-1}, but control samples without acetylene should always be tested for ethylene production.

(5) In *Beijerinckia indica* cultures the nitrogenase activity increases as the acetylene partial pressure is increased up to about 0.75 atm.[82] Most other N_2-fixing systems analysed have a K_m for C_2H_2 less than 0.01 atm.[37]

When we developed an assay for field soil samples some 4 years ago, we found that the more the soil was disturbed the lower was the activity and the greater the variability. Eventually we found that soil cores, 11 cm diameter × 16 cm long, retained in metal tubes after extraction from the field, provided reproducible samples for small plants. The core is placed in a glass bottle closed with a rubber septum, acetylene is added to a concentration of about 15% v/v and it is then left overnight. Pressure in the bottle is then released through a needle, and the first gas sample taken. Several hours later, another sample is taken to determine the rate of ethylene production. For larger samples assays can be carried out in plastics containers; these also allow some or all of the plant tops to be retained.

When the soil–plant state is disturbed, a variable time lag with low activity is induced before the nitrogenase activity resumes at a rate comparable with that of the undisturbed state. This lag can be up to 20 h for plant roots, regardless of the presence of acetylene or the oxygen tension used for the incubation.[28, 29, 39, 72] For roots and rhizomes extracted from the soil, the nitrogenase activity is dependent on the oxygen tension, with the greatest activity (nearest that of the intact system) usually occurring at a Po_2 of 0.04 atm or less for both tropical and temperate plants.[26, 28, 29, 39] For root pieces more reproducible results are obtained when they are moistened after removal from the soil, and pre-incubated for 16 h in a mixture of 4% oxygen in argon or nitrogen before adding acetylene.

For sandy soils an *in situ* assay system has been used. A 30 cm diam. metal tube is driven 5–7 cm into the ground, and the plant tops enclosed by a plastics bag or rigid plastics dome connected to the metal tube by a water seal. Acetylene is released inside the chamber by dropping water onto calcium carbide.[3] In the heavy clay soil at Rothamsted this was unsatisfactory because of the very slow penetration of the acetylene into the soil, beyond the top 12 cm.[21]

The *in situ* assay system has been modified by us to study N_2 fixation by blue-green algae on the soil surface. Metal tubes, 15 cm in diameter, are driven about 5 cm into the soil (with 2–3 cm projecting above the soil) some time before the assay and, if necessary, contact between soil and tube improved by adding some water to the interface. On the upper surface a clear Perspex plate containing a Suba Seal is sealed with a non-setting material. Acetylene is injected from a cylinder, to give about 20% v/v in the atmosphere above the soil surface. Diffusion of acetylene into Rothamsted clay soil is rapid for the first 20 min and then proceeds more slowly and almost linearly for several hours. Gas samples are taken 0.5 and 1 h after injection of the acetylene for measurement of ethylene production. The loss of ethylene through the soil is taken into account.[34, 94]

Suitable assay techniques have still to be developed for large tropical plants such as sorghum, millet and *Pennisetum purpureum,* the root systems of which are known to stimulate non-symbiotic N_2 fixation.

12.3 Nitrogenase activity in temperate regions

In Broadbalk Field

The Broadbalk experiment at Rothamsted has provided the most convincing evidence for non-symbiotic nitrogen fixation in temperate soils. In 1843 Lawes set up the Permanent Wheat Experiment on Broadbalk, a field with clay with flints soil. Previously the field had been limed, and the soil pH has subsequently remained at about 7. The field was divided into a series of plots manured in the same way each year. Several fertiliser combinations were used, including one with no nutrient additions (plot 3), one with P, K, Na, Mg additions but no N (plot 5), and others with increasing levels of N fertiliser. In 1966 the nitrogen content of the top 23 cm of plot 10, which had received by then 12 000 kg N per hectare as ammonium sulphate, was 0.106%, little greater than that of plot 3. The content of nitrogen in the grain removed and the soil has been monitored since the start of the experiment. After taking into account dry sorption of ammonia by the soil (calculated to be 13 kg ha^{-1} year^{-1}) and nitrogen added in the rainfall (measured at 5.4 kg ha^{-1} year^{-1}). the nitrogen balance for plots receiving no fertiliser nitrogen shows a steady rate of nitrogen gain of about 30 kg ha^{-1} year^{-1}.[47]

Leguminous and other weeds, mostly cluster clover and vetch, are too few to account for much of this fixation. Virtually no nitrogenase activity was found in the fallow soil and in the soil beneath the surface among the wheat plants. Assays of soil cores containing wheat plants showed that the amount of fixation associated with their roots was low and could account for only 2–3 kg ha^{-1} year^{-1}.[21] Blue-green algae on the soil surface, often forming a continuous crust, as Bristol Roach noted in 1927, may account for most of the fixation. Little nitrogenase activity was found before mid-May, and the activity varies considerably from season to season, depending on the rainfall and evaporation. Hence, activity in sections not treated with herbicide is often higher than in treated sections. In 1972 plot 6 (given 48 kg N ha^{-1} year^{-1}) was estimated to have most nitrogen fixation (28 kg N ha^{-1} year^{-1}). Plots 3 and 5, with less plant cover, so that the soil surface dried out more rapidly, had estimated fixations ranging from 8 to 23 kg N ha^{-1} year^{-1}. Sometimes nitrogenase activities were as high as 1–2 kg N ha^{-1} day^{-1}, especially after heavy rain late in the Summer. Nitrogenase activity continued into the night, with about 15% of the total daily activity occurring between 10 P.M. and 6 A.M. The species involved were mainly *Nostoc ellipsosporum* with *Nostoc punctiforme* and *Cylindrospermum* sp. also present. Inoculation of the soil with these algae in May increased activity slightly, mainly through earlier establishment of the algal cover.[34, 94]

In 1882 part of Broadbalk was enclosed and left uncultivated. Part of this 'Broadbalk Wilderness' has developed into mixed woodland with a sparse ground flora, now predominantly *Hedera helix* (ivy), with patches of *Mercurialis perennis* (dog's mercury) and isolated plants of *Heracleum sphondylium* (hogweed). Another part of the Wilderness has been stubbed annually to remove woody species, and now consists of coarse grasses and a mixed dicotyledonous flora of about 40 species (*Figure 12.1a*). Large amounts of nitrogen have steadily accumulated in both parts with a net gain in the top 20 cm of soil of more than 2000 kg N ha^{-1} since 1882. This represents a minimum annual gain from non-symbiotic fixation of 49 and 39 kg N ha^{-1} for the wooded and stubbed parts, respectively, because legumes have been virtually absent since about 1915 and no other nodulating plants are present.[46] In the Wilderness most of the nitrogenase activity was found closely associated with the roots of several species of dicotyledonous weeds from many genera (*Table 12.1*). Non-rhizosphere fixation in soil alone; phyllosphere fixation on leaves and fixation associated with bark and litter accounted for, at most, 4–5 kg N ha^{-1} year^{-1}.[39]

There was little difference between the nitrogenase activities in 11 cm diameter soil cores containing *Heracleum sphondylium*, *Nepita glechoma* or *Stachys sylvatica* plants, and activity was present to a depth of 75 cm, with fixation rates estimated at as much as 440 g N ha^{-1} day^{-1} on occasions. Activity was significantly correlated with soil moisture contents. Moist soils probably have a more favourable (i.e. lower) oxygen tension for fixation associated with the plant roots. The activity associated with plant roots in the stubbed section could account for the nitrogen gains over the years. On the wooded Wilderness the soil was quite dry during most of the 1973 sampling period, and activities associated with the ground cover plants were too low to account for the nitrogen accumulation. Perhaps most activity occurs in early Spring or late Autumn, when the trees have least effect on the soil moisture content, or some activity may be associated with tree roots.[21]

232

(a)

(c)

(b)

Figure 12.1 (a) Broadbalk Wilderness with the stubbed section in the foreground and the wooded section behind. The arable field lies to the left with the manurial strips running at right angles to the direction of viewing. (b) Paspalum notatum growing in sandy soil in experimental plots at IPEACS, near Rio de Janeiro. The plants in the right foreground are the cultivar Batatais, which has an active nitrogen-fixing root association; the cultivar Pensacola on the left has little associated fixation. (By courtesy of J. Döbereiner.) (c) Two Digitaria decumbens plants transplanted into the same pot of vermiculite watered with nitrogen-free nutrient solution. The large plant on the right has an active nitrogen-fixing association; the plant on the left has not. There is little or no transfer of the fixed nitrogen from the rhizosphere of the active plant to the other plant. (By courtesy of J. Döbereiner)

233

Table 12.1

PLANTS ON BROADBALK WILDERNESS
WITH MOST ASSOCIATED
NITROGENASE ACTIVITY

Stachys sylvatica	hedge woundwort
Mercurialis perennis	dog's mercury
Heracleum sphondylium	hogweed
Anthriscus sylvestris	cow parsley
Rumex acetosa	sorrel
Convolvulus arvensis	bindweed
Viola canina	dog's violet
Nepita glechoma	ground ivy
Hedera helix	ivy

The soil of Geescroft Wilderness nearby, started similarly in 1883, has a pH of between 4.2 and 4.8; analyses have shown non-symbiotic N_2 fixation to have been about 9 kg N ha^{-1} year^{-1}.[46] Nitrogenase activity when the soil was reasonably moist, under the dominant ground cover of *Hedera helix,* was similar to that of the Broadbalk wooded section, which suggests that pH is not a major factor limiting non-symbiotic nitrogen fixation.[21]

In other wooded areas

Nitrogen gains of about 50 kg N ha^{-1} year^{-1} have been recorded elsewhere in Britain for 11 monocultural non-nodulated tree stands over a period of 20 years[64] and for *Pinus* spp in Australia.[71] A *Pinus taeda* forest in South Carolina, USA, where the vegetation had been annually burned, accumulated 23 kg ha^{-1}year^{-1} over a 10 year period.[50] Small nitrogenase activities have been found associated with washed roots of several conifers[71, 80] and with roots and leaves of Douglas fir.[49]

Nitrogenase activity has also been detected in living white fir trees, associated with fungal attack,[77] and in rotting wood.[16, 78] Termites, which feed on wood, have nitrogen-fixing bacteria in their hindgut and these may be their main N-source. Termites play an important role in decomposing dead trees and other cellulose-containing materials, and may possibly add to the soil nitrogen supply by this fixation, particularly in tropical regions.[8, 13] Earthworms and their casts likewise show a small nitrogenase activity.

In non-wooded areas

Experiments with lysimeters have shown annual gains of nitrogen of 45 kg N ha^{-1} under mustard and 50–74 kg N ha^{-1} under perennial grasses and irises (see reference 62). In Western Australia a ryegrass pasture without legumes on fine soil accumulated 185 kg N ha^{-1} in 3 years. When such grass swards follow wheat crops, the fertility of the soils is regenerated, as revealed by subsequent crops; this may be largely due to non-symbiotic nitrogen fixation. This does not occur on coarser soils, perhaps because better aeration decreases nitrogen fixation.[65] In some habitats such as the Canadian prairies, with highly organic soils, only 2–3 kg N ha^{-1} is fixed annually.[67] Much of this activity is associated with mosses containing blue-green algae.[88] In peaty soils activity is also low.[90]

Experiments at Rothamsted, as early as 1917, and elsewhere showed that adding straw or farmyard manure to soil may increase nitrogen fixation, particularly under waterlogged conditions.[5, 15] Whether much fixation occurs when stubble is ploughed in is a moot point. During decomposition, straw often produces substances toxic to young cereal plants, and with continuous cereal cropping the possible benefits from increased nitrogen fixation have to be measured against such possible harmful effects.

Adding sugar or molasses to many soils greatly stimulates nitrogenase activity with little delay, which suggests that nitrogen-fixing bacteria may be limited by lack of carbohydrate nutrients (see e.g., reference 15).

Non-symbiotic nitrogen fixation is important in tundra areas. Nodulated plants do occur, although relatively sparsely. Appreciable nitrogenase activity occurs in the surface 10 cm of soil, even though the soil temperature rarely exceeds 5°C. Most activity was found in sedge-dominated, poorly drained meadows.[85] Mosses and lichens are also abundant in such regions and their associated blue-green algae—often intercellular—also fix nitrogen.[2]

Blue-green algae have been isolated from both Antarctic waters and hot thermal springs and so may well fix nitrogen over a range of temperatures from near 0°C to about 50°C.[83] Their ability to survive high temperatures and desiccation in the akinete form enables them to develop rapidly when the soil becomes moist. In desert conditions algae often form surface crusts, active in nitrogen fixation.[33, 57, 73] On Broadbalk Field mildly desiccated, surface crusts can show nitrogenase activity within 2 h of rewetting—a remarkable transformation from an inactive to an active state which suggests that some heterocysts and vegetative cells have survived desiccation.[94] Lichen associations of several genera of blue-green algae and fungi exist over a range of habitats from the arctic to the desert because of this ability; nitrogenase activity per unit of algal protein in such associations may be much higher than for algae in pure culture, and the activity depends on a rapid transfer of the fixed nitrogen from alga to fungus (e.g. references 33, 44, 52, 60). *Nostoc* also forms a symbiosis with *Gunnera* capable of fixing most of the nitrogen required by the plant.[7, 79]

Blue-green algae also have an important niche in less extreme conditions. They abound in estuaries and salt marshes, where they actively fix nitrogen (see references 33, 83). Lake water usually has a low concentration of inorganic nitrogen compounds, so favouring nitrogen-fixing organisms. When phosphorus concentrations are not limiting, blooms of nitrogen-fixing blue-green algae often result.[83, 86] Such algae are also widely distributed in non-acidic soils, where their activity is limited by complete plant cover, but in exposed, moist situations or in shallow ponds they fix nitrogen at a considerable rate.[35, 41, 42]

In sand dunes blue-green algae also play a part in nitrogen accumulation, the nitrogen fixed being transferred to associated plants.[84] Dune plants such as *Ammophila arenaria*[40] and *Spinifex hirsutus* (A. W. Moore, personal communication) also appear to be involved in this, possibly through fixation by rhizosphere-associated bacteria. In this habitat the low supply of nutrients is the major limiting factor to build-up of fertility.

Water plants can also stimulate nitrogen fixation by the bacteria tightly bound to their roots and rhizomes. The water cover provides an anaerobic environment favouring nitrogenase activity. The tropical sea grasses *Thalassia, Syringodium* and *Diplanthera* and the temperate *Zostera* have active associations in some localities (see, e.g., references 58, 66), with *Thalassia* stands estimated to fix enough to satisfy the plants' nitrogen requirements. Similarly, the fresh-water plants *Glyceria borealis* and *Typha* sp. stimulate nitrogen fixation, estimated for the *Glyceria* system to be as much as 60 kg N ha^{-1} year^{-1}.[14]

Thus, the two most important non-symbiotic nitrogen-fixing systems in temperate regions are the bacteria bound to the rhizosphere of several dicotyledonous and a few monocotyledonous plants, and blue-green algae on the soil surface, in water or inside plant cells.

12.4 Nitrogenase activity under tropical conditions

Many systems of tropical agriculture rely heavily on non-symbiotic nitrogen fixation for maintenance of soil fertility. In shifting agriculture the bush fallow in between periods of cultivation often contains legumes, but these are not usually dominant, which implies that non-symbiotic fixation is important in nitrogen accumulation during the fallow.[38, 63] In savannah regions nitrogen is lost during the frequent grass burning. When not under cultivation, grasses, not legumes, provide the dominant ground cover, although leguminous shrubs and trees may be present. Grain legume crops are often used in the rotation, but cereals and root crops predominate. Rice has been grown continuously on the same fields in Asia for perhaps thousands of years with little change in soil fertility although no nitrogenous fertiliser has been added. The nitrogen removed in the rice crop has been replaced by non-symbiotic nitrogen fixation.

Associated with grasses

Nitrogen-fixing bacteria have been found in many tropical soils. As many of these soils are acid, *Beijerinckia* and *Clostridium* sp. are the most common species found, with few *Azotobacter chroococcum*.[6, 25] Several monocotyledonous plants, particularly sugar cane, stimulate the multiplication of *Beijerinckia* spp. in their rhizospheres.[24] However, the unimproved, broad-leafed, cultivar 'batatais' of *Paspalum notatum* (*Figure 12.1b*), a grass which dominates many acidic, low-nitrogen-status, sandy soils in Brazil, specifically stimulates the nitrogen-fixing bacterium *Azotobacter paspali. A. paspali* establishes poorly in the rhizospheres of other cultivars and species.[25]

Work at Rothamsted showed that considerable nitrogenase activity was associated with the roots of batatais plants.[28] Root and rhizome pieces incubated aerobically had little nitrogenase activity, but at a Po_2 of around 0.04 atm activity was maintained for more than 24 h after a lag period of 12–20 h. Fifteen centimetre diameter soil cores containing plants showed no lag, and activity was similar either in air or at a Po_2 of 0.04 atm, which shows that the intact system provides a suitable oxygen tension for nitrogen fixation. Washing roots to remove the soil had little effect on activity. The activity of the intact core was greater than the sum of the activities obtained from the soil, root or leaf pieces incubated separately. It was estimated that *Paspalum notatum* could fix 100 kg N ha^{-1} year^{-1}. Starting from small

237

rhizome pieces, *P. notatum* plants grown with nitrogen-free nutrient solution in vermiculite fixed 84 mg N in 2 months.[26]

Sugar cane also stimulates much rhizosphere fixation but soil midway between rows containing no roots was also active, presumably because run-off from the leaves after rain is heaviest there, bringing sugar from the leaves to provide an energy source for fixation.[29]

Most of the important tropical grasses from several countries in three continents have now been found to have considerable nitrogenase activity in their rhizospheres (*Figure 12.1c* and *Table 12.2*).[4, 20, 22, 26, 28, 29, 51, 69, 72, 91, 95] Low activity is also found in the phyllosphere of several plants, particularly in the humid tropics, but is probably not important to the plant.[10, 74]

Root-associated activity is related to the growth stage of the plant with little after flowering and varies during the season and from season to season, probably in response to changes in soil moisture and temperature.[22] The fixation seems to be closely linked to photosynthesis, changing during the day and with peaks of activity for *P. notatum* and *Sorghum vulgare* near midday and near midnight.[26] The loss of activity when plants are kept in the dark is restored within a few hours of returning them to the light.[28] Adding sugar to roots and rhizomes of *Paspalum notatum* increased activity of plants kept in the dark, but not that of unshaded plants, which implies that a close relationship exists between numbers of nitrogen-fixing organisms and carbohydrate supply in the rhizosphere.[26] Adding sugar to non-rhizosphere soil obtained from under *P. notatum*[28] and several other grasses from Nigeria[20] increased nitrogenase activity without multiplication of nitrogen-fixing bacteria, which shows that activity in soil is limited by carbohydrate availability.

Experiments using $^{15}N_2$ showed that some fixed nitrogen was transferred from bacteria to plant within 17 h, which suggests that the bacteria excreted some nitrogen compound that was readily absorbed by the plant.[27] How much fixed nitrogen goes directly into the plant, rather than after decomposition of the bacteria, is not known. Assays of *Pennisetum purpureum* plants in the field suggested that nitrogen fixation supplied about two-thirds of the plant's nitrogen. With *P. purpureum, Digitaria decumbens* and *Paspalum notatum,* fertiliser additions of 20 kg N ha^{-1} every 2 weeks had no effect on plant growth even after eight additions, which shows that nitrogenase activity alone can meet the nitrogen demand of the plants. Nitrogenase activity decreases within 2 h of adding ammonium, but after 24 h both nitrate and ammonium fertilised

238

Table 12.2

N_2-FIXING ASSOCIATIONS WITH TROPICAL MONOCOTYLEDONS

Plant	Origin	$N_2 ase$ activity (expressed as nmoles C_2H_4/g root/h)
Brachiaria mutica	Brazil	150–750
B. rugulosa		5–150
Cymbopogon citratus		2
Cynodon dactylon		20–270
Cyperus rotundus		10–30
Digitaria decumbens		20–400
Hyparrhenia rufa		20–30
Melinis minutiflora		15–40
Panicum maximum		20–300
Paspalum notatum		2–300
Pennisetum purpureum		5–1 000
Saccharum officinarum (sugar cane)		5–20
Sorghum vulgare seedlings		10–100
Andropogon gayanus	Nigeria	15–270
Cenchrus ciliaris		16
Cymbopogon giganteus		60–85
Cynodon dactylon		10–50
Cyperus spp.		2
Hyparrhenia rufa		30–140
Hypothelia dissoluta		10–15
Panicum maximum		75
Paspalum commersonii		25–30
P. virgatum		3
Pennisetum purpureum		60
P. coloratum		13
Setaria anceps		1–120
Pennisetum typhoides (millet)		3–195
Sorghum		22–83
Andropogon spp.	Ivory Coast	50–380
Brachiaria brachylopha		100–140
Bulbostylis aphyllanthoides		74
Cyperus obtusiflorus		30–620
C. zollingeri		50–160
Cyperus sp.		150–1 900
Fimbristylis sp.		80–190
Hyparrhenia dissoluta		2–4
Loudetia simplex		54
Maize seedlings	France	100–3 000
	USA	14–16
Rice	Phillipines	8–80
Rice seedlings (in test-tubes)	Ivory Coast	1 040–2 360
	France	10–3 000
Setaria anceps	Australia	68
Pennisetum clandestinum		21–140

plants had similar low activities. Within a week activity is restored to that of unfertilised plants,[22] which suggests that nitrogenous fertiliser additions have only a transitory effect on this rhizosphere-associated nitrogen fixation.

Nitrogen fixation associated with *Paspalum notatum, Pennisetum purpureum*[26] and *Digitaria decumbens*[27] varied a great deal between cultivars and even ecotypes, which suggests that there is much scope for plant breeding in stimulating this activity (*Figure 12.1b*). *Paspalum notatum* cv. batatias roots have a thick mucilaginous coating in places, and this perhaps provides the right niche for fixation (*Figure 12.2a*). Activity is localised on the root system of field-grown *Digitaria decumbens* plants, with most activity on medium-thickness roots near sites of lateral root emergence, over half the root system inactive, and less than a third with much activity. Such active root parts have more cortical cells filled with bacteria than less active parts, which suggests that these may be concerned with nitrogen fixation.[27] This could explain why washing roots removes little activity. Such an association has similarities to the non-legume root nodule.

Associated with cereals

Some tropically grown cereals also stimulate nitrogenase activity, rice being the most active plant, with fixation by bacteria on the root surface.[51, 72, 95] Nitrogen-fixing blue-green algae are commonly found in rice fields.[89] Inoculating paddy water with such algae also stimulates growth and yield of rice even when given nitrogen fertiliser.[87] Part of this response may be due to gibberellin or auxin production by the algae. At present such inoculation is economically worthwhile. Photosynthetic, nitrogen-fixing bacteria also abound in rice fields in Japan, and are believed to add significantly to the nitrogen balance of such soils.[53] In Asia the small fern *Azolla* is abundant on the paddy water and soil surface and contains *Anabaena azollae* in the substomatal chamber; this association is estimated to fix up to 120 kg N ha^{-1} per season.[7, 33, 61] The benefit from this is appreciated by Vietnamese farmers, who inoculate rice fields with *Azolla* at the start of the season.

With the increasing cost of nitrogenous fertilisers and the increasing use of rice varieties responsive to fertilisers, the possibility of complementing fertiliser treatment with nitrogen fixation in rice culture becomes more important.

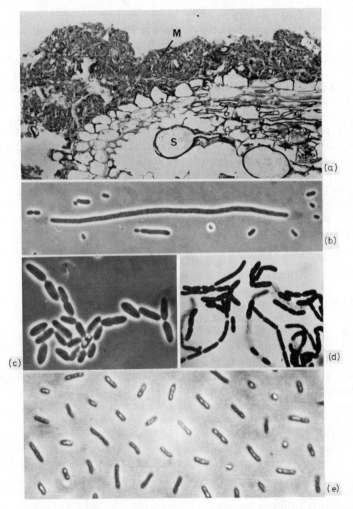

Figure 12.2 (a) Section of a Paspalum notatum *batatais root showing the extensive mucilaginous layer (M) coating the epidermis. The root also has an extensive* Endogone *mycorrhizal infection, with two spores (S) present in this section. ×144. (b)* Azotobacter paspali *from a nitrogen-free broth culture, showing the diversity of cell size. ×530. (c)* Azotobacter chroococcum *from nitrogen-free broth culture with little variation in cell size. ×960. (d) Gram-positive, spore-forming nitrogen-fixing aerobe, similar to* Bacillus polymyxa, *isolated from the rhizosphere of a Nigerian grass. ×1060. (e) Gram-negative nitrogen-fixing aerobe isolated from* Stachys sylvatica *roots, with characteristic inclusions (poly-β-hydroxybutyrate). ×1520.*

Nitrogenase activity has also been found on maize,[69, 91] sorghum and millet roots in the field (Jones and Dart, unpublished), and on roots of yam and sweet potato grown in pots.

All tropical plants so far found to stimulate much activity on their roots, except rice, possess the C-4 dicarboxylic acid photosynthetic pathway as opposed to the C-3 Calvin cycle of most temperate plants. Plants using the former pathway have a rapid rate of photosynthesis, which is not saturated by the high light intensities of the tropics and has a temperature optimum between 30 and 40°C. More dry matter is accumulated per unit of water transpired, and such plants have the highest growth rate in the world (see, e.g., reference 12). These plants are thought to translocate more of their carbohydrate to the roots and to exude more in their rhizospheres. It is interesting that the plant which was associated with nitrogen fixation in some of Berthelot's experiments in 1885, *Amaranthus pyramidalis,* is probably a plant of this type.

12.5 Organisms involved

Some 47 species of bacteria from 12 families have been shown to fix nitrogen. They are found in many habitats, including water, and represent a variety of Gram-negative and Gram-positive organisms, aerobic, facultative anaerobic and anaerobic, spore- and cyst-forming, photosynthetic and non-photosynthetic methane-oxidising and sulphate-reducing organisms.[18] Eighteen genera of blue-green algae are able to fix nitrogen.[18, 33] It is not yet clear which bacteria are most involved in fixing nitrogen in plant rhizospheres. Bacteria can be isolated from nitrogen-fixing root pieces, but many isolates are difficult to maintain in pure culture. Some grow better in mixed culture, associated with non-nitrogen-fixing organisms, perhaps a yeast or, as Cutler and Bal[17] showed in 1926, a protozoan. One has the impression that many of the organisms involved in nitrogen fixation on roots are fastidious in their nutritional requirements. Because of difficulties in isolating or culturing anaerobes, we know little of their involvement. Many of the bacteria isolated can grow anaerobically, but when grown in air or even a Po_2 of 0.04 atm, have a different colony form.

Organisms commonly isolated belong to the *Entero-bacteriaceae—Klebsiella, Enterobacter* or *Escherichia* species. Anomalous reactions in cultural tests make identification of species difficult. Such organisms have been found in high numbers on the roots of *Stachys sylvatica* and *Mercurialis perennis* at Rothamsted;[39] on maize and soy-bean plants or decaying wood in North America;[31, 77] on roots of several grasses from Northern Nigeria;[20] and on grass roots in New Zealand.[56] *Klebsiella* and *Enterobacter* species have large capsules, polysaccharide in nature, and much extracellular material, some organised in strands, which stain with Toluidine Blue or carbol Fuchsin, in a similar manner to proteins (*Figure 12.3e, f*). One role of such extracellular material may be to provide a favourable oxygen tension for fixation.

The organism *Derxia gummosa,* commonly found in tropical soils,[25] grows aerobically on agar, forming colonies with a hard surface.[43] We isolated a similar organism from earthworm casts on Broadbalk (*Figure 12.3a*). Such organisms grow with difficulty and may be overlooked if a dilution method is used.

Counting nitrogen-fixing bacteria on roots is difficult. Dilution counts become unreliable if the growth of the organisms is affected by the dilution procedure.[11] To overcome this, a method in which soil crumbs are sprinkled on nitrogen-free agar has been used. The most difficult problem, however, is to remove bacteria from the root surface quantitatively. If some are in root cortex cells, maceration is required. Some non-nitrogen-fixing bacteria grow well on minimal amounts of-combined nitrogen and confuse dilution plate counts. Finding a suitable medium is not easy. The important organism *Spirillum lipoferum* (*Figure 12.3b*), first isolated by Beijerinck in 1923, and found in both temperate[75] and tropical soils,[27] has been overlooked because of its special carbon requirements, having a preference for malate or aspartate. Glucose and sucrose are not used. It is interesting that malate and aspartate are early intermediates in the C-4 photosynthetic pathway, but are not formed in the C-3 pathway. If excess malate were transported to the roots, this might give a selective advantage to organisms such as *S. lipoferum*. To isolate *S. lipoferum* from the roots of *Digitaria decumbens* or other tropical grasses, a root piece is immersed in a semi-solid, nitrogen-free, malate medium, and after incubating for about 30 h a pellicle of growth occurs around the root piece containing almost a pure culture of the *Spirillum*. These crude cultures are very efficient in nitrogen fixation, fixing up to 40 mg N/g malate, as measured by Kjeldahl analysis.[27]

Beijerinckia indica (*Figure 12.3c*) and *Beijerinckia fluminensis* (*Figure 12.3d*) are common in tropical soils, and show a marked rhizosphere effect in some plants.[24, 25] *B. fluminensis* forms an envelope enclosing two or more bacteria, possibly as a means of

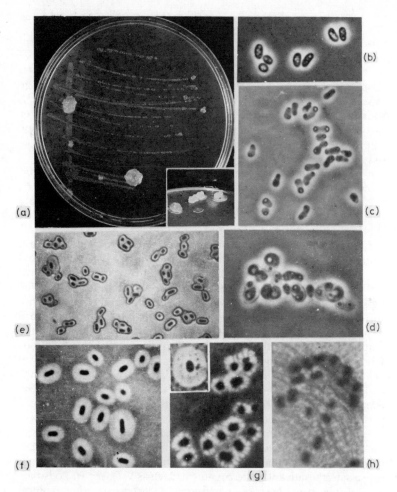

Figure 12.3 (a) Plate showing growth of a nitrogen-fixing culture isolated from earthworm casts from the Broadbalk Wilderness. Characteristically a few, isolated, massive colonies with a very tough surface develop. The inset shows another plate of the same organism, on which only these colonies developed. (b) Spirillum lipoferum. *(c)* Beijerinckia indica *with characteristic polar granules. (d)* Beijerinckia fluminensis *with two or more bacteria enclosed by an envelope. (e)* Enterobacter *isolated from* Stachys sylvatica *root, stained with periodic acid–Schiff's reagent. (f)–(h)* Enterobacter *isolated from Nigerian grass root, showing large capsule. (g) is stained with Toluidine Blue. Stained strands of material cross the capsule. Much material external to the capsule lies between the bacteria and stains a dense blue, which indicates a basophilic substance. Inset, Indian ink preparation. Note the dense material within the capsule. (h) Carbol Fuchsin stain showing strands of densely stained material radiating from the bacteria. (b)–(h) are all from nitrogen-free broth culture and × 1300. (b)–(d) are phase contrast, unstained*

244

protection from oxygen. *Klebsiella pneumoniae* isolated from *Stachys sylvatica* roots when stained with periodic acid–Schiff's reagent have a dense border to their capsules (*Figure 12.3e*) reminiscent of the *Beijerinckia fluminensis* envelope. A Gram-positive, spore-forming organism from Nigerian grass roots was similar to *Bacillus polymyxa* (*Figure 12.2d*).

Azotobacter are usually present in small numbers or not at all on root surfaces, even after inoculation. The plant-growth-stimulating effects obtained by inoculation with *Azotobacter* cultures are thought to be due to hormone production by the bacteria and not to nitrogen fixation (see Chapter 2). However, *Paspalum notatum* shows a marked rhizosphere effect on *Azotobacter paspali,* but numbers and nitrogen fixation are not well correlated.[28] Roots can be washed free of *A. paspali* yet still retain some nitrogenase activity (M. E. Brown, personal communication), which suggests that other organisms may be responsible for some fixation. For various plants *A. paspali* causes similar growth stimulating effects to *Azotobacter chroococcum.* *Azotobacter paspali* is a fascinating organism, producing cells in culture up to 150 μm long which bud off smaller ovoid cells similar in size to *Azotobacter chroococcum* (*Figure 12.2b, c*).

A possible exception to this diminished nitrogen-fixing role for *Azotobacter* occurs in the Near East. High numbers of *A. chroococcum* are reported for cultivated soils (up to 10^7/g soil) and also for irrigation water. Turning in plant residues enhances the numbers and increases nitrogen fixation. Nitrogen-fixing clostridia are also present in high numbers (up to 10^8/g soil).[1]

Although it is possible to establish nitrogen-fixing plant associations from plant cuttings, such a system has yet to be established with sterilised seedlings inoculated with pure bacterial cultures. There appears to be a widespread distribution of appropriate nitrogen-fixing organisms in soils, so that the prospect of benefit from inoculation of seeds is remote. The main limiting factor appears to be the availability of carbohydrate in the plant root.

12.6 Significance for agriculture

Lichens and blue-green algae are important as initial colonisers of rock surfaces in soil development. Many plant habitats can efficiently recycle their nitrogen, so that their requirements from N_2 fixation are low. Inputs from rainfall seem low (usually less than 10 kg N ha^{-1}

year^{-1}) in both temperate and tropical regions (see, e.g., reference 30). After burning of vegetation most, if not all, of the nitrogen lost is returned in the rain, albeit spread over a wider area than that burned. Disturbing these natural systems by burning or cropping, as in shifting agriculture, increases the amount of nitrogen required to maintain equilibrium. The actual inputs are usually greater when nodulated plants are a component of the vegetation.

In temperate regions non-symbiotic microbial nitrogen fixation is unimportant where high crop yields demand massive inputs of fertiliser nitrogen. Under the present systems of tillage for cereals the open soil structure perhaps militates against non-symbiotic nitrogen fixation associated with plant roots. The use of herbicides also reduces a potential nitrogen source by killing weeds. If straw were turned in rather than burned, non-symbiotic nitrogen fixation would probably be increased in arable fields. Only in woodland and rough grazing areas does non-symbiotic nitrogen fixation appear to have a significant role. At present we are ignorant of the potential of temperate grasses to stimulate nitrogen fixation in their rhizospheres.

In temperate regions with extensive agricultural systems nitrogen fixation can balance the relatively low productivity in some soils. Wheat production in parts of Australia has involved rotations with few or no legumes with little decline in yields. Introducing pasture or undersown legumes into the rotations, however, greatly increases productivity.

For tropical soils the contribution of non-symbiotic nitrogen fixation is much greater. Less than 10% of the non-arable land in South America and Africa is improved pasture.[32] In the natural vegetation legumes are usually sparse. However, calculations show that nitrogen fixation, mostly by non-symbiotic organisms, can be considerable, with gains of up to 165 kg N ha^{-1} year^{-1} for lowland forests, 73 kg for regenerated rain forest, 45 kg for highland forest and 17 kg for savannah.[36] Field experiments over 3 years at Ibadan, Nigeria, in the lowland humid tropic zone showed accumulations of 595 kg N ha^{-1} year^{-1} for regenerated bush without legumes, and 90 kg under the grass *Cynodon plectostachyus*.[45] Extrapolations from our acetylene reduction assays indicate that other tropical grasses are capable of greater levels of nitrogen fixation. Grasses stimulating fixation also have a high rate of dry-matter accumulation and this is a valuable adjunct to legumes in pastures. Fortunately, some of the improved grasses, such as *Setaria anceps* cv. Kazangula and Nandi, and cultivars of *Pennisetum purpureum*, *Pennisetum clandestinum* and

246

Digitaria decumbens, are very active in stimulating N_2 fixation and at the same time more digestible by animals than unselected, coarse, native grasses. Blue-green algae also fix much more under tropical than temperate conditions, where the soil moisture may limit algal growth (see, e.g., reference 33). The non-symbiotic fixation associated with cereals is potentially the most valuable for tropical agriculture. Several field experiments with sorghum and millet (and also sugar cane) showed little or no response to nitrogen fertilising, and calculations of nitrogen balances together with nitrogenase assays suggest that the root-associated non-symbiotic fixation is contributing much of their nitrogen. Maize, a plant with the C-4 photosynthetic pathway, has been extensively bred, and perhaps this accounts for the relatively low nitrogen fixation levels measured to date in temperate regions; possibly the higher temperatures of the tropics would stimulate more activity. Breeding plants for increased nitrogen fixation has exciting possibilities. To date we know little of the effect of other factors such as soil moisture, fertility levels and cutting frequency for grasses on non-symbiotic nitrogen fixation in plant rhizospheres.

Unfolding the potential for non-symbiotic nitrogen fixation in maintaining soil fertility promises to be a rewarding area for research.

REFERENCES

1. ABD-EL-MALEK, Y. Free living nitrogen-fixing bacteria in Egyptian soils and their possible contribution to soil fertility, *Plant and Soil Special Volume,* 423–442 (1971)
2. ALEXANDER, V. A. Nitrogen fixation by algae in polar and sub-polar regions, in *Nitrogen Fixation by Free-living Microorganisms* (edited by W. D. P. Stewart). 179–188, Cambridge University Press (1975)
3. BALANDREAU, J. and DOMMERGUES, Y. Assaying nitrogenase (C_2H_2) activity in the field, *Bulletins from the Ecological Research Committee,* **17,** 247–254 (1973)
4. BALANDREAU, J. and VILLEMIN, G. Fixation biologique de l'azote moléculaire en savane de lamto (basse cote d'Ivoire) résultats préliminaires, *Revue d'Écologie et de Biologie du Sol,* **10,** 26–33 (1973)
5. BARROW, N. J. and JENKINSON, D. S. The effect of water-logging on fixation of nitrogen by soil incubated with straw, *Plant and Soil,* **16,** 258–262 (1962)
6. BECKING, J. H. Studies on nitrogen fixing bacteria of the genus Beijerinkia. 1. Geographical and ecological distribution in soils, *Plant and Soil,* **14,** 49–81 (1961)
7. BECKING, J. H. Nitrogen fixation in some natural ecosystems in Indonesia, in *Symbiotic Nitrogen Fixation* (edited by P. S. Nutman). Cambridge University Press (in press)
8. BENEMANN, J. R. Nitrogen fixation in termites, *Science,* **181,** 164–165 (1973)

Non-symbiotic Nitrogen Fixation in Soil

9. BERTHELOT, M. Fixation directe de l'azote atmosphérique libre par certains terrains argileux, *Compte Rendu (hébdomadiare) des séances de l'Academie du Sciences,* **101,** 775–784 (1885)

10. BESSEMS, E. P. M. Nitrogen fixation in the phyllosphere of Gramineae, *Agricultural Research Reports.* Centre for Agricultural Publishing and Documentation, Wageningen, 786 (1973)

11. BILLSON, S., WILLIAMS, K. and POSTGATE, J. R. A note on the effect of diluents on determination of viable numbers of Azotobacteriaceae, *Journal of Applied Bacteriology,* **33,** 270–273 (1970)

12. BLACK, C. C. Ecological implications of dividing plants into groups with distinct photosynthetic production capacities, *Advances in Ecological Research,* **7,** 87–114 (1971)

13. BREZNAK, J. A., BRILL, W. J., MERTINS, J. W. and COPPEL, H. C. Nitrogen fixation in termites, *Nature,* **244,** 577–580 (1973)

14. BRISTOW, J. M. Nitrogen fixation in the rhizosphere of freshwater angiosperms, *Canadian Journal of Botany,* **52,** 217–221 (1974)

15. BROUZES, R., MAYFIELD, C. I. and KNOWLES, R. Effect of oxygen partial pressure on nitrogen fixation and acetylene reduction in a sandy loam soil amended with glucose, *Plant and Soil Special Volume,* 481–494 (1971)

16. CORNABY, B. W. and WAIDE, J. B. Nitrogen fixation in decaying chesnut logs, *Plant and Soil,* **39,** 445–448 (1973)

17. CUTLER, D. W. and BAL, D. V. Influence of protozoa on the process of nitrogen fixation by *Azotobacter chroococcum, Annals of Applied Biology,* **13,** 516–534 (1926)

18. DALTON, H. Fixation of dinitrogen by free-living microorganisms, *CRC Critical Reviews in Microbiology,* **3,** 183–220 (1974)

19. DART, P. J., DAY, J. M. and HARRIS, D. Assay of nitrogenase activity by acetylene reduction, in *Use of Isotopes for Study of Fertiliser Utilization by Legume Crops,* 85–100. International Atomic Energy Agency, Vienna (1972)

20. DART, P. J., HARRIS, D. and DAY, J. M. Nitrogen fixation associated with the roots of tropical grasses, *Report of Rothamsted Experimental Station for 1972, Part 1,* 87–88 (1973)

21. DAY, J., HARRIS, D., DART, P. J. and VAN BERKUM, P. The Broadbalk experiment. An investigation into gains from non-symbiotic nitrogen fixation, in *Nitrogen Fixation by Free-living Micro-organisms* (edited by W. D. P. Stewart). 71–84, Cambridge University Press (1975)

22. DAY, J. M., NEVES, M. C. P. and DÖBEREINER, J. Nitrogen fixation on the roots of tropical forage grasses, *Soil Biology and Biochemistry* (in press)

23. DILWORTH, M. J. Acetylene reduction by nitrogen-fixing preparations from *Clostridium pasteurianum, Biochimica et Biophysica Acta,* **127,** 285–294 (1966)

24. DOBEREINER, J. Nitrogen-fixing bacteria of the genus *Beijerinckia* Derx in the rhizosphere of sugar cane, *Plant and Soil,* **15,** 211–216 (1961)

25. DÖBERINER, J. and CAMPELO, A. B. Non-symbiotic nitrogen fixing bacteria in tropical soils, *Plant and Soil Special Volume,* 457–470 (1971)

26. DÖBEREINER, J. and DAY, J. M. Nitrogen fixation in the rhizosphere of tropical grasses, in *Nitrogen Fixation by Free-living Micro-organisms* (edited by W. D. P. Stewart). 39–56, Cambridge University Press (1975)

27. DÖBEREINER, J. and DAY, J. M. Associative symbioses in tropical grasses: characterization of microorganisms and dinitrogen fixing sites, in *International Symposium on Nitrogen Fixation,* Washington State University (in press)

28. DÖBEREINER, J., DAY, J. M. and DART, P. J. Nitrogenase activity and oxygen sensitivity of the *Paspalum notatum—Azotobacter paspali* association, *Journal of General Microbiology,* **71,** 103–116 (1972)

29. DÖBEREINER, J., DAY, J. M. and DART, P. J. Nitrogenase activity in the rhizosphere of sugar cane and some other tropical grasses, *Plant and Soil*, **37,** 191–196 (1972)

30. ERIKSSON, E. Composition of atmospheric precipitation. 1. Nitrogen compounds, *Tellus*, **4,** 145–270 (1952)

31. EVANS, H. I., CAMPBELL, N. E. R. and HILL, S. Asymbiotic nitrogen-fixing bacteria from the surface of nodulues and roots of legumes, *Canadian Journal of Microbiology*, **18,** 13–21 (1972)

32. *FAO Production Yearbook*, 24 (1970)

33. FOGG, D. E., STEWART, W. D. P., FAY, P. and WALSBY, A. E. *The Blue-green Algae*. Academic Press, London (1973)

34. FROGGAT, P. J., KEAY, P. J., WITTY, J. F., DART, P. J. and DAY, J. M. Algal fixation in Rothamsted Fields, *Report of Rothamsted Experimental Station for 1972, Part 1*, 87 (1973)

35. GRANHALL, U. Studies on nitrogen fixation by algae in temperate soils, in *Nitrogen Fixation by Free-living Micro-organisms* (edited by W. D. P. Stewart). 189–198. Cambridge University Press (1975)

36. GREENLAND, D. J. and NYE, P. H. Increases in the carbon and nitrogen contents of tropical soils under natural fallows, *Journal of Soil Science*, **10,** 284–299 (1959)

37. HARDY, R. W. F., BURNS, R. C. and HOLSTEN, R. D. Applications of the acetylene–ethylene assay for measurement of nitrogen-fixation, *Soil Biology and Biochemistry*, **5,** 47–81 (1973)

38. HARRIS, D. R. The ecology of Swidden cultivation in the upper Orinoco rain forest Venezuela, *Geographical Review*, **61,** 475–495 (1971)

39. HARRIS, D. and DART, P. J. Nitrogenase activity in the rhizosphere of *Stachys sylvatica* and some other dicotyledenous plants, *Soil Biology and Biochemistry*, **5,** 277–279 (1973)

40. HASSOUNA, M. G. and WAREING, P. F. Possible role of rhizosphere bacteria in the nitrogen nutrition of *Ammophila arenaria*, *Nature*, **202,** 467–469 (1964)

41. HENRIKSSON, E., ENGLUND, B., HEDÉN, M. B. and WAS, I. Nitrogen fixation in Swedish soils by blue green algae, in *Taxonomy and Biology of Blue Green Algae* (edited by T. V. Desikachary), 269–273. Centre for Advanced Study in Botany, Madras (1972)

42. HENRIKSSON, E. Algal nitrogen fixation in temperate regions, *Plant and Soil Special Volume*, 415–419 (1971)

43. HILL, S. Influence of oxygen concentration on the colony type of *Derxia gummosa* grown on nitrogen-free media, *Journal of General Microbiology*, **67,** 77–83 (1971)

44. HITCH, C. J. B. and STEWART, W. D. P. Nitrogen fixation by lichens in Scotland, *New Phytologist*, **72,** 509–524 (1973)

45. JAIYEBO, E. O. and MOORE, A. W. Soil nitrogen accretion under different covers in a tropical rain-forest environment, *Nature*, **197,** 317–318 (1963)

46. JENKINSON, D. S. The accumulation of organic matter in soil left uncultivated, *Report of Rothamsted Experimental Station for 1970, Part 2*, 113–137 (1971)

47. JENKINSON, D. S. Organic matter and nitrogen in soils of the Rothamsted Classical Experiments, *Journal of the Science of Food and Agriculture*, **24,** 1149–1150 (1973)

48. JENSEN, H. L. Contribution to the nitrogen economy of Australian wheat soils, with particular reference to New South Wales, *Proceedings of the Linnean Society of New South Wales*, **65,** 1–22 (1940)

49. JONES, K. Nitrogen fixation in the phyllosphere of the Douglas Fir, *Pseudotsuga douglasii*, *Annals of Botany*, **34,** 239–244 (1970)

50. JORGENSEN, J. R. and WELLS, C. G. Apparent nitrogen fixation in soil influenced by prescribed burning, *Soil Science Society of America Proceedings*, **35,** 806–810 (1971)

Non-symbiotic Nitrogen Fixation in Soil

51. KALININSKAYA, T. A., RAO, V. R., VOLKOVA, T. N. and IPPOLITOV, L. T. Determination of the nitrogen-fixing activity of soil collected under rice plantings by acetylene reduction assay, *Microbiologiya*, **42**, 481–485 (1973)

52. KALLIO, P., SUHONEN, S. and KALLIO, H. The ecology of nitrogen fixation in *Nephroma arcticum* and *Solorina crocea*, *Report Kevo Subarctic Research Station*, **9**, 7–14 (1972)

53. KOBAYASHI, M. and HAQUE, M. Z. Contribution to nitrogen fixation and soil fertility by photosynthetic bacteria, *Plant and Soil Special Volume*, 443–456 (1971)

54. LAWES, J. B. and GILBERT, J. H. On the present position of the question of the nitrogen of vegetation with some new results and preliminary notice of new lines of investigation, *Philosophical Transactions of the Royal Society of London*, **180**, 1–107 (1889)

55. LAWES, J. B., GILBERT, J. H. and PUGH, E. On the sources of nitrogen of vegetation; with special reference to the question of whether plants assimilate free or uncombined nitrogen, *Philosophical Transactions of the Royal Society of London*, **151**, 431–577 (1861)

56. LINE, M. A. and LOUTIT, M. W. Non-symbiotic nitrogen-fixing organisms from some New Zealand tussock grassland soils, *Journal of General Microbiology*, **66**, 309–318 (1971)

57. MAYLAND, H. F., McINTOSH, T. H. and FULLER, W. H. Fixation of isotopic nitrogen on a semi-arid soil by algal crust organisms, *Soil Science Society of America Proceedings*, **30**, 56–60 (1966)

58. MCROY, C. P., GOERING, J. J. and CHANEY, B. Nitrogen fixation associated with sea grasses, *Limnology and Oceanography*, **18**, 998–1002 (1973)

59. MEIKLEJOHN, J. Azotobacter numbers on Broadbalk Field, Rothamsted, *Plant and Soil*, **23**, 227–235 (1965)

60. MILLBANK, J. W. Nitrogen metabolism in Lichens. IV. The nitrogenase activity of the *Nostoc* phycobiont in *Peltigera canina*, *New Phytologist*, **72**, 1–10 (1972)

61. MOORE, A. W. Azolla: Biology and Agronomic significance, *Botanical Review*, **35**, 17–34 (1969)

62. MOORE, A. W. Non-symbiotic nitrogen fixation in soil and soil plant systems, *Soils and Fertilizers*, **29**, 113–128 (1966)

63. NYE, P. H. and GREENLAND, D. J. The soil under shifting cultivation, *Technical Communication No. 51, Commonwealth Bureau of Soils* (1960)

64. OVINGTON, J. D. Studies of the development of woodland conditions under different trees. IV. The ignition loss, water, carbon and nitrogen content of the mineral soil, *Journal of Ecology*, **44**, 171–179 (1956)

65. PARKER, C. A. Non-symbiotic nitrogen-fixing bacteria in soil. III. Total nitrogen changes in a field soil, *Journal of Soil Science*, **8**, 48–59 (1957)

66. PATRIQUIN, D. and KNOWLES, R. Nitrogen fixation in the rhizosphere of marine angiosperms, *Marine Biology*, **16**, 49–58 (1972)

67. PAUL, E. A., MYERS, R. J. K. and RICE, W. A. Nitrogen fixation in grassland and associated cultivated ecosystems, *Plant and Soil Special Volume*, 495–507 (1971)

68. PROKSCH, G. Application of mass- and emission-spectrometry for $^{14}N/^{15}N$ ratio determination in biological material, in *Isotopes and Radiation in Soil–Plant Relationships, Including Forestry*, 217–225. IAEA, Vienna (1972)

69. RAJU, P. N., EVANS, H. J. and SEIDLER, R. J. An asymbiotic nitrogen-fixing bacterium from the root environment of corn, *Proceedings of the National Academy of Science of the United States of America*, **69**, 3474–3478 (1972)

70. RICE, W. A. and PAUL, E. A. The acetylene reduction assay for measuring nitrogen fixation in waterlogged soil, *Canadian Journal of Microbiology*, **17**, 1049–1056 (1971)

71. RICHARDS, B. N. Nitrogen fixation in the rhizosphere of conifers, *Soil Biology and Biochemistry*, **5**, 149–152 (1973)

72. RINAUDO, G., BALANDREAU, J. and DOMMERGUES, Y. Algal and bacterial non-symbiotic nitrogen fixation in paddy soils, *Plant and Soil Special Volume*, 471–479 (1971)

73. ROGERS, R. W., LANGE, R. T. and NICHOLAS, D. J. D. Nitrogen fixation by lichens of arid soil crusts, *Nature*, **209**, 96–97 (1966)

74. RUINEN, J. The phyllosphere. V. The grass sheath, a habitat for nitrogen-fixing micro-organisms, *Plant and Soil*, **33**, 661–671 (1970)

75. SCHRÖDER, M. Die assimilation des Luftstickstoffs durch einige Bakterien, *Zentralblatt für Bakteriologie, Parasitenkunde Infektionskrankheiten und Hygiene*, Abt. II, **85**, 177–212 (1932)

76. SCHÖLLHORN, R. and BURRIS, R. H. Acetylene as a competitive inhibitor of N_2 fixation, *Proceedings of the National Academy of Sciences of the United States of America*, **58**, 213–216 (1967)

77. SEIDLER, R. J., AHO, P. E., RAJU, P. N. and EVANS, H. J. Nitrogen fixation by bacterial isolates from decay in living white fir trees (*Abies concolor* (Govd and Glend) Lindl.), *Journal of General Microbiology*, **73**, 413–416 (1972)

78. SHARP, R. F. and MILLBANK, J. W. Nitrogen fixation in deteriorating wood, *Experientia*. **29**, 895–896 (1973)

79. SILVESTER, W. B. Endophyte adaptation in *Gunnera–Nostoc* symbiosis, in *Symbiotic Nitrogen Fixation* (edited by P. S. Nutman). Cambridge University Press (in press)

80. SILVESTER, W. B. and BENNET, K. J. Acetylene reduction by roots and associated soil of New Zealand conifers, *Soil Biology and Biochemistry*, **5**, 171–179 (1973)

81. SMITH, K. A. and RESTALL, S. W. F. The occurrence of ethylene in anaerobic soil, *Journal of Soil Science*, **22**, 430–443 (1971)

82. SPIFF, E. D. and ODU, C. T. I. Acetylene reduction by *Beijerinckia* under various partial pressures of oxygen and acetylene, *Journal of General Microbiology*, **78**, 207–209 (1973)

83. STEWART, W. D. P. Nitrogen fixation by photosynthetic microorganisms, *Annual Review of Microbiology*, **27**, 283–316 (1973)

84. STEWART, W. D. P. Transfer of biologically fixed nitrogen in a sand dune slack region, *Nature*, **214**, 603–604 (1967)

85. STUTZ, R. C. and BLISS, L. C. Acetylene reduction assay for nitrogen fixation under field conditions in remote areas, *Plant and Soil*, **38**, 209–213 (1973)

86. VANDERHOEF, L. N., DANA, B., ENNERICH, D. and BURRIS, R. H. Acetylene reduction in relation to levels of phosphate and fixed nitrogen in Green Bay, *New Phytologist*, **71**, 1097–1105 (1972)

87. VENKATARAMAN, G. S. *Algal Biofertilizers and Rice Cultivation*. Today & Tomorrow's Printers & Publishers, New Delhi (1972)

88. VLASSAK, K., PAUL, E. A. and HARRIS, R. E. Assessment of biological nitrogen fixation in grassland and associated sites, *Plant and Soil*, **38**, 637–649 (1973)

89. WATANABE, A. and YAMAMOTO, Y. Algal nitrogen fixation in the tropics, *Plant and Soil Special Volume*, 403–413 (1971)

90. WAUGHMAN, G. J. and BELLAMY, D. J. Acetylene reduction in surface peat, *Oikos*, **23**, 353–358 (1972)

91. WEINHARD, P., BALANDREAU, J., RINAUDO, G. and DOMMERGUES, Y. Fixation non symbiotique de l'azote dans la rhizosphère de quelques non-légumineuses tropicales, *Révue d'Écologie et de Biologie du sol*, **8**, 367–373 (1971)

92. WILSON, P. W. On the sources of nitrogen of vegetation etc., *Bacteriological Reviews*, **21**, 215–226 (1957)

251

Non-symbiotic Nitrogen Fixation in Soil

93. WILSON, P. W. Asymbiotic nitrogen fixation, in *Encyclopaedia of Plant Physiology*, Vol. 8 (edited by W. Ruhland), 9–47. Springer-Verlag, Berlin (1958)
94. WITTY, J. F. Algal nitrogen fixation on solid surfaces and in temperate agricultural soils, PhD thesis, University of London (1974)
95. YOSHIDA, T. and ANCAJAS, R. R. Nitrogen fixing activity in upland and flooded rice fields, *Soil Science Society of America Proceedings*, **37**, 42–46 (1973)

Index

253

Index

Index

Index

Medium, for ammonia oxidisers, 135
Melinis minutiflora, phosphorus uptake by, 85
Meristem, apical, 39, 40
Mesopores, 152
'Meta' fission, 184
Metabolic pathways, 188, 189
Metabolite theory, 215
Methane-forming bacteria, 9
Microbial degradations of pesticides, examples of, 186
Micropores, 152
Mineralisation of soil nitrogen, 208, 210
Moisture, effect of, on soil protozoa, 154, 158
Molecules, recalcitrant, 183
Molybdenum content
    of plants, effect of rhizosphere microflora on, 28
    of vegetables, relationship with dental caries, 28
Monochlorbenzoates, cometabolism of, 189
Monochlorophenols, cometabolism of, 189
Monuron, utilisation of, as carbon source, 186
Mucigel
    definition of, 51
    ecological and physiological significance of, 25
    observation of, 25
    structure of, 25
    study of, 52
Mucilage
    boundary of, 51
    droplets, secretion of, 51
    electron micrographs of, 50, 51
    function of, in plants, 51
    layer, 39, 49–51
Mucilaginous droplets, at tip of maize roots, 50
Mucor, colonisation of steamed soil by, 199
Mycelial strands, 170
Mycorrhiza, 82–85
    beech, 83
    ectotrophic, 82
    inoculation with, 85
    sheathing, 173–177
    VA, 62, 83–85

Mycorrhizal fungi, ectotrophic, 41
Mycorrhizal plants, 82, 84
Mycorrhizal root, 83, 84
Mycorrhizal symbiosis, 86

N-Serve inhibitor, use of, 143
Nematodes, effect of soil fumigants on numbers of, 200
Nitrate
    and ammonium, balance between, 209
    in waterlogged soil, 11, 12
    increase in, 209, 210
    loss through leaching, 209
    reductase activity, 4
    reduction of, 3
    respiration, 3, 4
Nitrification
    effect of irradiation on, 209
    inhibition of, 208, 209
    inhibitors of, 12
Nitrite, oxidation of hydroxylamine to, 141
Nitrite oxidisers, 134, 136, 140
Nitrobacter, 136, 137, 140–142
Nitrococcus mobilis, description of, 140
Nitrogen
    ammonium, 207
    availability, 23
    fertiliser, loss of, by volatilisation, 27
    mineral, content of soil, 207
    mineralisation of, 207–209
    in seed exudates, 23
Nitrogen fixation
    correlation with polysaccharide production, 95
    by strains of soil clostridia, 7, 8
Nitrogenase activity, 227–232, 234, 235, 237, 238
    favouring of, by restricted aeration, 8, 9
Nitrosococcus, 137, 138
Nitrosococcus nitrosus, 138
Nitrosocystis, 134, 137, 138, 142
Nitrosocystis coccoides, 137, 138
Nitrosocystis oceanus, 138, 142
Nitrosogloea, 137
Nitrosomonas, 134, 136–139, 141, 142
Nitrosomonas europaea, 138, 139, 216
    hydroxylamine oxidase from, 142

258

259

Rhizomorphs, as mode of fungal growth, 170

Rhizosphere
calcium uptake in, 77
coincidence of, with zone of phosphate depletion, 73, 77
microbial population of, 77
organisms, phosphatase production by, 80
root exudates in, 77
soil, phosphate-decomposing organisms in, 79

Root cap, formation of film over, 52

Root cap cells, 39, 40, 45, 47
development of mucilage in, 50

Root exudates, as microbial substrate in rhizosphere, 77

Root hairs
bacterial entry through, 55
development of, 42, 43
nutrient uptake by, 42
secretion of mucilage droplets from, 51

Root organ cultures, sloughed cells in, 47

Root surface
of *Ammophila arenaria,* 49
of clover, 47
of strawberry, 47

Roots
beaded, 59
ontogeny of, 39
structure of, 39

Rothamsted soils
fermentative facultative anaerobes in, 4
Nitrobacter spp. isolated from, 140

Rumex acetosa, orthophosphate concentrations suitable for, 72

Russell and Hutchinson hypothesis, 147–149

Russula, genus, evolution of, 174

Saccharolytic forms, of soil clostridia, 6, 7, 14

Scabiosa columbaris, orthophosphate concentration suitable for, 72

Sea-water, nitrite-oxidising bacteria from, 140

Serratia marcescens, extracellular material obtained from, 97

Sheathing mycorrhizas, 173–176

Silica gel medium, 134, 136

Simultaneous adaptation hypothesis, 189

Size, of soil protozoa, 156, 157

Sloughed root cells, 39, 45, 47, 61

Soil aggregates, stabilisation of, 100–105

Soil
clostridia population of, 5
fauna of, effect of fumigation on, 200
fertility, improvement in, 147
fumigated, fungal recolonisation of, 199
gases, work on, 10
moisture content of, relationship of, with nitrogenase activity, 232
obligate anaerobes of, 4
pesticide degradation in, 185, 186
population, changes in, 195
pores, 152–155
Saxmundham, improvement in structure of, 13
steamed, fungal recolonisation of, 199
toxicity of, to plant growth, 211
water potential, 171–173
waterlogged, 11, 12

Soils
Broadbalk nitrogen-fixing clostridia in, 8
clostridia in, 4–7
nitrification in, 143
Rothamsted, fermentative facultative anaerobes in, 4

Solubilisation, of phosphate, 74, 75, 77, 78, 86

Sphaeroblasts, formation of in cortex of some stems, 57, 58

Spirillum lipoferum, isolation of, 243

Sporophores, loss of ability to produce, 175

Stabilisation, of soil aggregates, 100–105

Steamed soil
fungal recolonisation of, 199
protozoal content of, 200

Steaming, effect of, on nitrification, 208, 209

Sterilisation, of soil, 193

Straw
addition of, to soil, 78